HVAC BIBLE
FOR BEGINNERS

Complete Diy Guide to Master Any Type of Heating, Ventilation and Air Conditioning System | Ideal for Aspiring HVAC Pros without Any Prior Knowledge

Jason Mitchell

© **Copyright 2023: Jason Mitchell - All rights reserved.**

This document is geared towards providing exact and reliable information regarding the topic and issue covered. The publication is sold with the idea that the publisher is not required to render accounting, officially permitted or otherwise qualified services. If advice is necessary, legal or professional, a practiced individual in the profession should be ordered.
- From a Declaration of Principles, which was accepted and approved equally by a Committee of the American Bar Association and a Committee of Publishers and Associations.
In no way is it legal to reproduce, duplicate, or transmit any part of this document in either electronic means or printed format. Recording of this publication is strictly prohibited, and any storage of this document is not allowed unless with written permission from the publisher. All rights reserved.
The information provided herein is stated to be truthful and consistent in that any liability, in terms of inattention or otherwise, by any usage or abuse of any policies, processes, or directions contained within is the solitary and utter responsibility of the recipient reader. Under no circumstances will any legal responsibility or blame be held against the publisher for any reparation, damages, or monetary loss due to the information herein, either directly or indirectly.
Respective authors own all copyrights not held by the publisher.
The information herein is offered for informational purposes solely and is universal as so. The presentation of the information is without a contract or any type of guaranteed assurance.
The trademarks that are used are without any consent, and the publication of the trademark is without permission or backing by the trademark owner. All trademarks and brands within this book are for clarifying purposes only and are owned by the owners themselves, not affiliated with this document.

Table of Content

Part I HVAC fundamentals

Chapter 1
Introduction to HVAC 10

What is HVAC? 10
    HVAC Subsystems 10
    How HVAC Systems Work Together 11
    HVAC Distribution 12
    HVAC System Types 12
    HVAC Specialty Systems 13
    HVAC System Selection Factors 13

Brief History and Evolution 13
    Early Origins 14
    The Industrial Era 14
    The Birth of Modern HVAC 14
    Pioneers Who Advanced HVAC 15
    Ongoing Evolution 15

Importance of HVAC Systems 15
    Thermal Comfort 16
    Health and Wellness 16
    Productivity and Learning 17
    Efficiency and Sustainability 17
    Quality of Life and Economic Growth 17

Chapter 2
Basic Principles 18

Thermodynamics 18
    Key Thermodynamic Concepts 18
    Heat Transfer Methods 19
    Heating Fundamentals 19
    Cooling and Refrigeration 20
    Humidity Control 20
    Ventilation and Airflow 20
    Thermodynamic Mastery 20

Heat Transfer 21
    Conduction 21
    Convection 21

  Radiation 21

 Fluid Dynamics 22
  Key Principles 22
  Airflow 22
  Hydronic Systems 23
  Applications 23
  Emerging Trends 23

Chapter 3
Types of HVAC Systems 24

 Heating Systems 24
  Common Residential Systems 24
  Commercial and Industrial Systems 25

 Cooling Systems 25
 Ventilation Systems 27
  Why Is HVAC Ventilation Important? 27
  Key Components of Commercial HVAC Ventilation Systems 27
  Ventilation Design Considerations 28
  Common Ventilation System Types 28
  Proper Maintenance Ensures Ongoing Performance 29

 Hybrid Systems 29
  The Concept of HVAC Hybridization 30
  Realizing Significant Energy and Cost Savings 30
  Key Design Considerations for Success 30
  Evolving System Combinations and Integration 31
  An Optimistic Future for Sustainability 31

Part II Residential HVAC

Chapter 4
Components and Design 34

 Thermostats 34
  Types of Thermostats 34
  Working Mechanism 35
  Importance in HVAC Systems 35

 Ductwork 35
  Types of Ductworks 35
  Design Considerations 36
  Maintenance and Cleaning 37

 Filters and Air Quality 38
  Types of Filters 38
  Importance of Filters 39
  Filter Replacement and Maintenance 39

Chapter 5
Installation and Maintenance 41

DIY vs. Professional Installation 41
- The DIY Approach: Pros and Cons 41
- The Professional Installation Option 42
- Making the Choice 43

Routine Maintenance 43
Basic Troubleshooting 44
- General Troubleshooting Approach 44
- Common Troubleshooting Tips 44
- When to Seek Professional Help 44

Chapter 6
Energy Efficiency in Residential HVAC 45

Energy-saving Tips 45
- Regular Maintenance 45
- Proper Insulation and Sealing 46
- Programmable Thermostats 46
- Zoning Systems 47
- Energy-efficient Windows and Doors 47
- Utilize Natural Ventilation 47

Smart Thermostats 48
- Understanding Smart Thermostats 48
- Energy Reports and Insights 48
- Compatibility and Installation 48

Renewable Energy Options 49
- Solar Power 49
- Geothermal Heating and Cooling 49
- Wind Power 50

Part III Commercial HVAC

Chapter 7
Scale and Complexity 52

Differences Between Residential and Commercial 52
- Scale and Purpose 52
- Design Complexity and Installation 53
- Regulatory Environment 53
- Energy Efficiency and Sustainability 53
- Maintenance and Service 53

Equipment and Components 54
- Heating and Cooling Units 54
- Ductwork and Vents 54
- Thermostats and Controls 54
- Filters and Air Quality Components 55

Zoning and Controls 55
    Zoning 55
    Control Systems 55
    Energy Management 56
    Maintenance and Monitoring 56
    Building Automation Systems (BAS) 56
    Challenges in Commercial Zoning and Controls 57

Chapter 8
Specialized Applications 58

HVAC for Healthcare 58
    The Importance of HVAC in Healthcare 58
    HVAC System Components for Healthcare 59
    Energy Efficiency and Sustainability in Healthcare HVAC 59

HVAC for Manufacturing 60
HVAC in Office Spaces 60

Chapter 9
Commercial HVAC Maintenance and Regulations 62

Preventative Maintenance 62
    Importance of Preventative Maintenance 63
    Components of Preventative Maintenance 63
    Benefits of Preventative Maintenance 64

Energy Management Systems 64
    Understanding Energy Management Systems (EMS) 64
    Components of Energy Management Systems 65
    Benefits of Energy Management Systems 65
    Challenges and Considerations 66

Legal and Regulatory Guidelines 66
    Regulatory Bodies 66
    Compliance and Penalties 67
    Importance of Compliance 67

Part IV Advanced Topics and Technologies

Chapter 10
Emerging Trends 69

Smart HVAC Systems 69
    Components and Architecture 69
    Benefits 70
    Challenges 71

IoT in HVAC: Revolutionizing Climate Control 71
    Integration and Functionality 71
    Advantages 72
    Challenges 73

Future Predictions 73
   Green HVAC Technologies 73
   Artificial Intelligence and Machine Learning.. 74
   Modular and Flexible Designs 74
   Electrification and Heat Pumps 74
   Human-Centric Approach 75

Chapter 11
Troubleshooting and Repairs 76

Advanced Diagnostic Tools 76
   Instrumentation for Physical Systems... 76
   Software-Based Diagnostic Tools 77
   Importance of Data Analysis 78
When to Call a Professional 78
   Complex Systems and Specialized Knowledge 79
   Safety and Legal Concerns 79
   Persistent or Recurring Issues 79
   Time and Resource Constraints.. 79
   Cost-Effectiveness and Long-Term Benefits 80
   Maintaining Peace of Mind 80
Case Studies 80
   Case Study 1: Network Connectivity in a Corporate Environment................................ 80
   Case Study 2: Malfunctioning Industrial Robot in a Manufacturing Plant 81

Chapter 12
Sustainability and HVAC 83

Environmentally Friendly Practices 83
LEED Certification 84
Sustainable Technologies 85

Part V Practical Guide and Resources

Chapter 13
Choosing an HVAC Contractor 88

Questions to Ask... 88
   Experience and Expertise 88
   Licensing and Certification.. 89
   References and Reviews 89
   Insurance Coverage.. 89
   Warranty and Guarantees 89
   Energy Efficiency and Environmental Concerns 89
   Project Timeline and Schedule 90
   Payment Terms and Pricing 90
   Communication and Accessibility 90
   Maintenance and Service Agreements 90

Verifying Credentials 90
    Licensing Verification 91
    Check for Complaints and Disciplinary Actions 91
    Insurance Validation 91
    Contact References 91
    Research Online Reviews 91

Getting Estimates 91
    On-Site Evaluation 92
    Detailed Written Estimate 92
    Compare Multiple Estimates 92
    Ask About Potential Additional Costs 92
    Clarify Payment Terms 92
    Review Contract Terms 92

Chapter 14
Cost Considerations 94

Budgeting for an HVAC System 94
    Understanding the Importance of HVAC Budgeting 94
    Factors Affecting HVAC Budgeting 95
    Creating an HVAC Budget 95
    Cost-Saving Strategies 96

Financing Options 97
    The Role of Financing in HVAC Projects 97
    Common HVAC Financing Options 97
    Choosing the Right Financing Option 97
    Planning for Successful HVAC Financing 98

Rebates and Tax Credits 98
    Types of HVAC Rebates and Tax Credits 99
    Understanding Eligibility and Application Processes 99
    Incorporating Incentives into the Budget 99
    Leveraging Incentives for Cost Efficiency 100

Conclusion 101
Claim Your Free Bonus 102

PART I
HVAC FUNDAMENTALS

CHAPTER 1
Introduction to HVAC

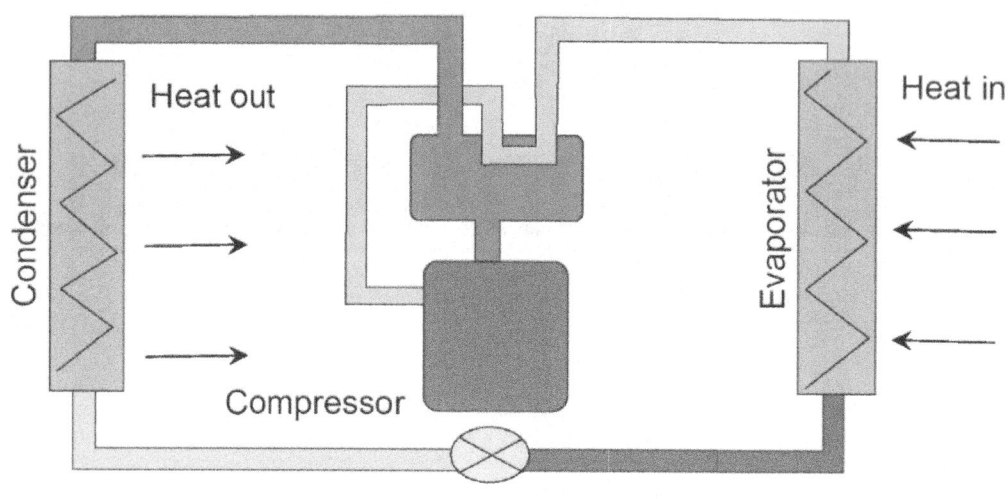

What is HVAC?

HVAC, or heating, ventilation, and air conditioning, refers to the systems and technology used to regulate temperature, humidity, airflow, and air quality in the spaces we occupy. Though often overlooked, HVAC is vital to our comfort, health, productivity, and quality of life.

From the ancient Romans channeling hot air through hollow floors to Willis Carrier's invention of modern air conditioning in 1902, humans have long sought control over their indoor environments. Today, HVAC is a $90 billion industry employing over 1.3 million people in the U.S. alone. The reach of these behind-the-scenes technologies extends into every home, business, factory, and institution.

But what exactly does HVAC entail? In broad terms, HVAC consists of the equipment, distribution systems, controls, and protocols used to heat, cool, ventilate, humidify, dehumidify, clean, and propel the air within a building. Designed thoughtfully, HVAC enables indoor climates tailored to human comfort and function.

HVAC Subsystems

HVAC operates through the integrated use of numerous types of equipment and distribution components.

Heating

Heating introduces thermal energy into space to raise temperatures. HVAC leverages various heat sources to warm air, water, or steam via:

- Furnaces - Utilize combustion of natural gas, propane, or fuel oil
- Boilers - Heat water to steam that is piped to heat exchangers
- Electric heating coils - Convert electricity into thermal energy
- Heat pumps - Move heat between indoor/outdoor air
- Solar thermal systems - Harness heat from the sun

Warm air or liquid is then distributed throughout buildings via ducts or pipes. Radiators, vents, radiant floors, and other delivery systems release warmth.

Cooling

Cooling does the opposite, removing thermal energy to lower temperatures. Popular equipment includes:

- Air conditioners - Refrigerant cooling cycles extract heat from indoor air
- Chillers - Machine systems that cool water sent to coil units
- Cooling towers - Disperse heat from water to air in industrial systems
- Evaporative coolers - Employ evaporation to lower air temps

Cold air is circulated through ducting to provide localized or centralized cooling.

Ventilation

Ventilation exchanges stale indoor air with fresh outdoor air. Methods include:

- Windows - Simple ventilation capable of fully exchanging room air
- Fans - Move air mechanically without conditioning
- Air handlers - Fan units that facilitate air distribution
- ERVs/HRVs - Heat recovery ventilators transfer energy between air streams

Controlled ventilation improves air quality and oxygenation.

Humidity Control

Ideal humidity levels for comfort and health range from 30% to 50%. HVAC regulates moisture through:

- Humidifiers - Add moisture to dry air via steam, mist, or evaporation
- Dehumidifiers - Condense and remove excess moisture
- Air conditioning - Cools air below the dew point, condensing moisture

Humidity control prevents problems like mold, condensation, and stuffiness.

Air Cleaning

Air filters, UV lamps, electrostatic precipitation, and other methods remove allergens and pollutants for healthier indoor air.

How HVAC Systems Work Together

Within a single building, multiple HVAC subsystems work in unison to create ideal conditions.

Sophisticated controls modulate different equipment as conditions change. Thermostats, humidistats, timers, and building automation systems monitor and adjust operations based on sensor data, settings, and usage patterns.

For example, a commercial HVAC setup may integrate chilled water cooling, forced air handlers, steam boilers, humidity control, and air cleaning to maintain office conditions through changing outdoor weather and occupancy.

Control systems ensure optimal efficiency by operating equipment sequencing. Heat and cooling are not simultaneously activated, and systems modulate to match needs.

HVAC Distribution

Conditioned air and fluids must be delivered to points of use. This is accomplished via:

Ductwork - Networks of sheet metal ducts deliver hot/cold air. Design factors include proper sizing, minimal leaks, room for maintenance access, and noise reduction.

Pipes and Pumps - For hydronic systems, pipes carry heated or chilled water to air handling units. Pumps circulate fluids.

Vents and Registers - Louvered vents, registers, grilles, and diffusers allow air to enter/exit the distribution system for heating, cooling, and ventilation.

Proper HVAC distribution ensures uniform conditions. Insulation reduces thermal losses en route. Good system design minimizes noise and improves indoor aesthetics.

HVAC System Types

There are two primary HVAC system categories: centralized and decentralized.

Centralized HVAC

Central systems consist of one or more large heating/cooling plants linked to a distribution network. This allows consolidated equipment to serve multiple zones from a central plant room or rooftop unit.

For example, a packaged rooftop unit may cool air that feeds ductwork spanning different floors of an office building. Centralized systems feature:

- Economy of scale for equipment.
- Single points of maintenance access.
- Wide temperature and humidity control across all zones.
- Challenges in maintaining comfort in diverse spaces.
- High installation costs.

Decentralized HVAC

Decentralized systems use multiple smaller HVAC units located in or near the zones they serve. This provides tailored conditioning for localized spaces such as:

- Multi-split systems - One condensing unit links to air handlers in different rooms.
- Variable refrigerant flow (VRF) systems - Refrigerant piping networks feed multiple evaporators.
- Unitary systems - Self-contained units for single zones.

Benefits include:

- Personalized comfort and efficiency.
- Incremental scalability.
- Reduced ductwork.
- Serviceability of individual units.
- Higher comparative costs for large facilities.

Each approach has advantages contingent on building and usage type.

HVAC Specialty Systems

In institutional, commercial, and industrial settings, HVAC often incorporates advanced systems like:

Laboratory ventilation - High air change rates, hazardous gas control, room pressurization.

Cleanrooms - Ultra-filtered, tightly controlled HVAC for contaminant control.

Data center cooling - Redundant high-capacity AC for heat-intensive computing.

Process cooling/heating - For manufacturing, chemical handling, food production, etc.

High-rise buildings - Zoned equipment stacks with height.

Specialized HVAC allows unique functions not possible through standard systems.

HVAC System Selection Factors

HVAC must suit unique needs. Key selection factors include:

- Usage - Commercial, residential, industrial, etc.
- Location/climate
- Budget
- Aesthetics
- Occupancy and spatial layout
- Single versus multi-zone needs
- System configuration preferences
- Available fuel types, electricity, water
- Integration complexity with existing systems
- Codes and regulations

Careful requirements analysis and design help build high-performing HVAC.

Proper heating, cooling, ventilation, humidity control, and air cleaning are vital to human habitats. HVAC systems represent the holistic orchestration of mechanical engineering in service of comfort and function. As energy-intensive infrastructure, improving efficiency and sustainability remain top priorities for the industry.

Brief History and Evolution

The drive to control indoor environments has motivated centuries of heating, ventilation, and air conditioning innovation. Each new advancement brought greater climate control and comfort - from ancient Rome's primitive HVAC to today's sophisticated systems.

Understanding this long history provides crucial context, allowing us to appreciate how modern heating and cooling conveniences emerged from the ingenuity of past civilizations.

Early Origins

The earliest forms of HVAC could hardly be called systems at all. Prehistoric humans gathered around fires, adapted their clothing and constructed basic shelters to mediate outdoor conditions. Over time, civilizations centralized these efforts, devising communal HVAC solutions to improve survival and comfort.

In Ancient Rome, the wealthy pioneered the West's first centralized heating by piping hot air through hollow floors and walls to warm baths and villas. Windows allowed natural ventilation, while water evaporation from fountains and damp cloths offered cooling.

In Medieval Europe, early Middle Ages heating reverted to simpler fireplaces and braziers. But later, master masons built enormous furnace-fed fireplaces spanning multiple floors - forerunners of central heating.

In Asia, Koreans perfected ondol underfloor heating, circulating smoke, and hot air through flues beneath stone floors. This concept spread, adapted as the Chinese kang and Japanese kotatsu. In the Americas, Native Americans constructed warm underground homes called hogans with centralized smoke holes for ventilation.

The Industrial Era

The 18th and 19th centuries saw great leaps in HVAC capabilities alongside industrialization. In the early 1800s, small low-pressure steam engines provided mechanical heating and ventilation. Steam gained broader use, powering large 19th-century factories. The American engineer Robert Mills pioneered central fan-driven ventilation, allowing climate control for larger buildings.

In the 1830s, the U.S. Capitol and other structures adopted European-style hot water radiant heating. On the cooling front, in the 1820s, British scientists Michael Faraday and John Dalton established principles enabling modern refrigeration. In 1842, Dr. John Gorrie patented an ice-making machine to cool hospital rooms.

Coal served as the predominant heating fuel through the mid-1900s. Cleaner burning oil and gas heating gained traction by the 1950s. New HVAC challenges also emerged around heating, cooling, and ventilating soaring urban high-rises, driving major innovation. And as reliable electricity spread in the late 1800s and early 1900s, electric heating and cooling equipment gained adoption.

The Birth of Modern HVAC

Early 20th-century breakthroughs led to the familiar HVAC forms we know today. In 1902, engineer Willis Carrier designed the world's first modern electric air conditioning system, launching an industry. Forced air heating furnaces transitioned from gravity-based to electrically powered blower systems.

The roaring 1920s accelerated HVAC adoption in theaters, offices, hotels, and upscale homes. Despite the 1929 financial crash, the 1930s still saw continued HVAC innovation, with more affordable residential systems emerging.

The postwar economic boom following WWII brought soaring commercial HVAC usage. Innovations like heat pumps, variable refrigerant flow (VRF) systems, and modern boiler controls also emerged. Analog electromechanical HVAC controls transitioned to digital for greater precision, birthing building automation systems.

With the 1970s energy crisis, efficiency became a top HVAC priority, accelerating innovation in high-performance equipment. After the 1980s epidemic of "sick building syndrome," HVAC paid greater attention to filtering and air change rates. As climate change impacts grew, sustainable HVAC improved energy efficiency and adopted cleaner refrigerants.

Pioneers Who Advanced HVAC

Certain pioneers made breakthroughs that greatly propelled HVAC capabilities forward. Willis Carrier called the "Father of Air Conditioning", founded Carrier Corporation and installed the first modern electric AC system. Engineer William Furness designed modern ductwork and ventilation principles in the early 20th century. Nikola Tesla's contributions to electricity and AC power helped drive electrified HVAC.

German engineer Gustav Zeuner created the first comprehensive thermodynamic theory for mechanical refrigeration. Edwin Ruud pioneered residential automatic gas water heaters and founded Ruud Manufacturing. Frederick McKinley Jones, the first African-American member of the American Society of Heating, Refrigerating, and Air-Conditioning Engineers (ASHRAE), innovated refrigerated trucking. Margaret Ingels, America's first female air conditioning engineer, consulted on the Pentagon's HVAC system.

Willis Carrier developed psychrometric charts precisely relating temperature, humidity, and other parameters to design effective HVAC systems. These charts are still used today.

Ongoing Evolution

Today's HVAC capabilities result from centuries of invention, turning revolutionary ideas into real systems heating, cooling, and ventilating the built world. Ongoing innovation promises even greater efficiency, sustainability, controllability, and integration with modern architecture and needs. By understanding the long history of HVAC, we better appreciate how these unseen technologies support human health, productivity, and comfort in the present day.

The drive of civilizations to continually improve their control over indoor environments fueled this progression. Each generation is built upon the HVAC advances of their forebears, bringing humanity closer to the heating, cooling, and ventilation capabilities we now depend on.

Importance of HVAC Systems

HVAC - is often overlooked yet vital to modern life. Heating, ventilation, and air conditioning form the technological foundation for built environments tailored to human comfort and function. By regulating temperature, humidity, airflow, and air quality indoors, HVAC enables spaces suitable for living, working, healing, learning, and all facets of human habitation.

Engineered thoughtfully, HVAC systems create indoor climates optimized for health, productivity, and wellbeing. They heat us through frigid winters, cool us during sweltering summers, refresh stale air, and adapt to changing weather and occupancy. Modern civilizations depend on HVAC to mediate between external environments and desired conditions inside.

This indoor climate control facilitates modern life itself. Without it, where would advanced civilizations operate? HVAC underpins industry, technology, medicine, culture – any human endeavor reliant on controlled spaces. Its essential status can be appreciated most when it fails. Equipment breakdowns, power outages, or disrupted heating fuel access quickly highlight our dependence. Buildings become unbearable and even dangerous, absent proper heating and cooling.

Thermal Comfort

An HVAC system's purpose is to establish and maintain human thermal comfort. Thermal comfort results from balanced heat gain and loss to hold the body's core temperature within a safe narrow range. Heat flows between individuals and their environments based on factors like air temperature, humidity, airflow, clothing, and activity levels. When heat loss exceeds gain, the body cools. The reverse causes overheating. Both threaten health if beyond tolerable deviations.

HVAC technology acts as a controlled interface managing heat exchange. Proper design considers heating/cooling equipment capacity, distribution, control schemes, and the building envelope. Maximum human thermal comfort is the goal.

Specialized facilities like hospitals, manufacturing plants, and data centers require precise conditions suited to their use. Parameters are adjusted accordingly. But for typical commercial and residential settings, general guidelines apply:

- Air temperatures between 68-75°F.
- Relative humidity 30-50%.
- Fresh outdoor air ventilation to dilute pollutants.
- Air distribution facilitates uniform conditions.
- Individualized local control.

Balancing these variables promotes both comfort and efficient operation.

Health and Wellness

In addition to comfort, properly controlled HVAC improves occupational health, safety, and wellbeing. Heating, cooling, humidity, and ventilation each have direct wellness impacts:

Heating

Protects from cold exposure injuries like hypothermia. Also wards off aggravating conditions like arthritis, Raynaud's syndrome, and chilblains.

Cooling

Prevents heat stress and associated dangers like heat stroke, exhaustion, rashes, cramps, or dizziness.

Humidity Control

Regulating moisture prevents mold/mildew and associated respiratory issues. Dry air can worsen conditions such as asthma.

Ventilation

Dilutes and removes indoor air pollutants, including volatile organic compounds (VOCs), bio-effluents, viruses, carbon dioxide, and other contaminants.

Research affirms this link between HVAC parameters and health. Cornell University studies found doubling ventilation rates reduced illness-related school absences by 10% to 20%. Indoor humidity control lowered flu transmission by up to 50%.

Thus, properly operating HVAC does more than simply make indoor conditions comfortable. The technologies fundamentally safeguard occupant health through all seasons.

Productivity and Learning

Comfort fuels productivity, learning, and performance. Environments that are too hot, cold, stuffy, or stale wear down mental acuity and physical endurance. Focus drifts to discomfort rather than tasks. Motivation suffers. Service and work quality decline.

School learning suffers without proper HVAC. A 2012 Harvard study showed classroom ventilation improvements raised test scores by 15-20 percent. Another found temperature extremes lowered scores up to 10 percent. The benefits extend to all workplaces. Comfortable conditions keep workers and operations humming.

Efficiency and Sustainability

HVAC accounts for roughly 40 percent of an average building's energy use. Total HVAC energy consumption represents over one-tenth of all energy used in the United States. This substantial level means HVAC offers major efficiency improvement opportunities.

Upgrading to modern high-efficiency equipment and optimizing control schemes can reduce HVAC energy usage by 30 percent or more. Swapping fossil fuels for electricity even enables deep decarbonization utilizing renewable energy. Efficient HVAC also cuts operating costs significantly over outdated systems.

These savings multiply across homes and businesses. More sustainable HVAC lowers national energy demand, carbon emissions, and reliance on imported fuels. Environmentally aware system selection and operation benefit both individual budgets and the planet.

Quality of Life and Economic Growth

Ultimately, the engineered comfort provided by HVAC systems enhances quality of life and enables economic growth. Global productivity depends on controlled indoor workplaces. Pharmaceuticals, electronics, and countless goods are manufactured within precise HVAC environments. Perishable food supplies stay fresh thanks to refrigerated cold chains. The internet hums inside temperature-regulated data centers.

Entire skyscraper cities would be uninhabitable without advanced HVAC capabilities. Comfortable homes raise living standards and expectations. Medical sterility depends on hospital HVAC. Museums protect delicate artifacts in climate-controlled exhibit halls. HVAC brings immense if subtle, societal value.

CHAPTER 2
Basic Principles

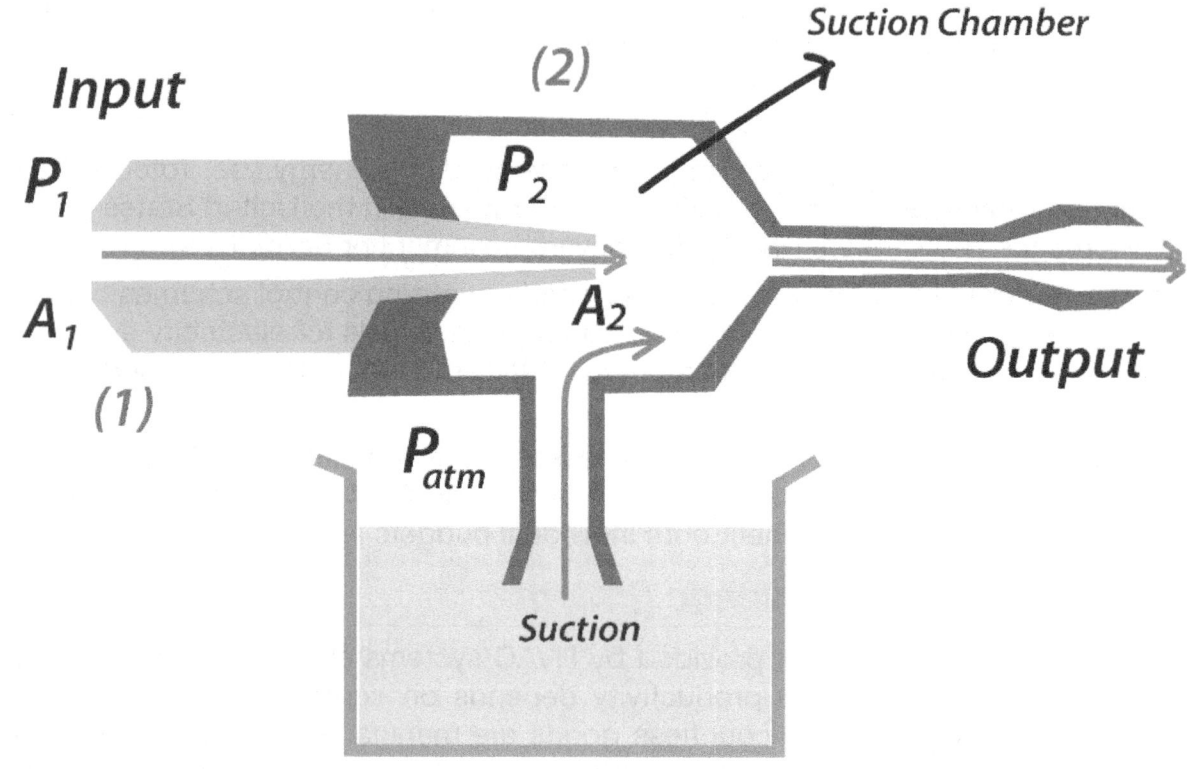

Thermodynamics

Underpinning all HVAC technology are the fundamental scientific principles of thermodynamics. These laws govern how heating, cooling, and all thermal energy transfers occur. By leveraging thermodynamics, HVAC systems regulate temperature, humidity, and mechanical work to achieve heating, ventilation, and air conditioning.

Thermodynamics emerged in the 18th and 19th centuries through the pioneering work of scientists like James Joule, Lord Kelvin, Rudolf Clausius, and Sadi Carnot. Their observations and experiments established how heat and work are related, how heat naturally flows, and the concept of energy in a closed system.

These basic rules provide HVAC engineers with invaluable understanding. All HVAC processes boil down to strategically managing thermodynamic energy flows. Whether heating an indoor space or generating chilled water, the equipment merely facilitates natural thermal and fluid dynamics. Mastering these basics is essential to effective HVAC design and operation.

Key Thermodynamic Concepts

HVAC systems leverage fundamental laws of thermodynamics to heat, cool, and ventilate spaces. Grasping a few core concepts provides insight into how these systems function:

Heat and Temperature

While related, heat and temperature differ crucially. Heat refers to the total amount of thermal energy present in a substance. Temperature measures the intensity or concentration of heat energy. Heating and cooling involve increasing or decreasing the total heat content of air.

States of Matter

Solids, liquids, and gases each store and transfer heat differently based on molecular structure. HVAC systems harness phase changes between states, like condensing/evaporating refrigerants.

Energy Transfer

Thermal energy spontaneously flows from higher to lower temperatures until equilibrium is reached. HVAC equipment manipulates the direction and rate of heat transfer through temperature differentials.

Work and Heat

In thermodynamics, work means using energy to induce temperature differences by adding or removing heat. HVAC components perform heating/cooling work by altering heat content.

Conservation of Energy

The first law of thermodynamics states energy cannot be created or destroyed, only changed in form. HVAC obeys this fundamental conservation of energy.

Entropy

Entropy quantifies the irreversible dispersal of energy during heat transfers. It always increases in a closed system. Efficient HVAC minimizes wasteful high-entropy heat flows.

Heat Transfer Methods

HVAC systems rely on the three modes of heat transfer - conduction, convection, and radiation - to heat and cool interior spaces.

Conduction is the transfer of thermal energy between objects in direct contact, such as heat traveling through a furnace wall. HVAC systems utilize conductive materials like metals to transport and radiate heat efficiently. Convection involves the fluid movement of heat through liquids and gases like air. HVAC ductwork leverages the convective motion of forced air to circulate heated and cooled air streams throughout a building. Radiation is the emission of infrared energy directly through space without the need for conduction or convection. Radiant HVAC utilizes this electromagnetic radiation, allowing surfaces like ceilings and floors to heat spaces.

By maximizing the heat transfer modes, which efficiently convey thermal energy where needed for climate control while minimizing undesired modes like heat gain/loss from the environment, HVAC equipment regulates interior comfort and temperature.

Heating Fundamentals

All HVAC heating involves raising heat content by adding thermal energy to air or liquid. This uses thermodynamic concepts:

- Adding heat raises temperature (q+).

- Heat naturally flows from higher to lower temperatures.
- Heat input does mechanical work by driving the temperature change.
- Energy cannot be created or destroyed, only transferred.

For example, a gas furnace combusts fuel to heat an air stream. The higher temperature air is circulated through ducts and convectively transfers warmth to cool indoor spaces. The total heat always remains constant, just shifting location.

Cooling and Refrigeration

Cooling removes heat using sequential thermodynamic effects:

First, a refrigerant absorbs heat from its surroundings in an evaporator, changing from a liquid to a gas state. This cools the evaporator's vicinity.

The now-heated gas is then mechanically compressed to raise its temperature higher than outdoor air.

The hot gas passes through a condenser, transferring its heat outside. This condenses the refrigerant back into a liquid, which can repeat the cycle.

This clever "refrigeration cycle" uses heat flow principles to create cooling. The same thermo rules govern air conditioners, fridges, chillers, and all HVAC refrigeration.

Humidity Control

Humidity, or moisture level, relates closely to thermodynamic psychrometrics - quantifying air-water vapor mixtures. Control schemes leverage phase change dynamics between liquid water and gaseous water vapor.

Dehumidification condenses excess vapor into liquid water, which is drained away using refrigeration equipment and surface chillers. Humidification adds water vapor through steam injection, misting, and evaporative systems.

By managing moisture content, HVAC can maintain both comfort and indoor air quality.

Ventilation and Airflow

Air movement for ventilation and convection relies on fluid dynamics - another branch of physics underpinning HVAC. Fans generate air pressure gradients to circulate fresh, oxygenated air. Ducts route airflow, while diffusers induce desired patterns through space.

Automating ventilation rates based on occupancy and sensors helps ensure adequate indoor air quality.

Thermodynamic Mastery

From combustion furnaces to vapor compression chillers, all HVAC systems leverage underlying thermodynamic transfers at the particle level. Mastering the fundamentals of heat, temperature, energy, change of state, and heat flow empowers HVAC engineers to manipulate these forces for the benefit of human comfort.

Combined with fluid dynamics governing ventilation and humidity, thermodynamics provides the scientific foundation for the technologies that keep our built environments liveable year-round. Physics and HVAC are intimately intertwined.

Heat Transfer

Understanding how heat moves is crucial for anyone working in HVAC. Heat transfer, also called heat flux, is the movement of thermal energy from one object, region, or substance to another. This occurs due to a temperature difference between the objects.

There are three main mechanisms through which heat is transferred - conduction, convection, and radiation. Let's take a closer look at each one to understand how heat moves through different materials and surroundings.

Conduction

Conduction is the transfer of heat through direct contact between objects or substances. When two objects at different temperatures are in direct contact, thermal energy flows from the warmer object to the cooler one until thermal equilibrium is reached. During conduction, energy is transferred by the physical interaction of neighboring atoms and molecules within a material.

The rate of heat transfer by conduction depends on several factors, such as the temperature gradient or difference between the objects, the area of contact between them, and the thermal conductivity of the material. Metals are excellent conductors of heat as they have a high density of free electrons, which allows thermal energy to flow rapidly through the material. On the other hand, plastics, wood, glass, and other non-metals are poor conductors as their molecules are tightly packed with limited mobility.

Convection

While conduction requires physical contact between objects, convection relies on the actual movement of a fluid like air, water, or other liquids for heat transfer. As the temperature of the fluid increases, its molecules gain kinetic energy and move faster. This causes the density of the warmer parts to decrease compared to the cooler parts. The difference in density results in buoyancy, which causes vertical movement in fluids and the creation of convection currents.

For example, when a pot of water is heated on a stove, the water at the bottom nearest to the heating element gets warmer and less dense than the surrounding water. This low-density hot water rises upward while the higher-density cooler water sinks downward, setting up a convection current. Such currents aid efficient and rapid heat transfer throughout the fluid. Air, too, exhibits convection, which is why heating and cooling systems rely on circulating airflow inside rooms.

Radiation

Unlike conduction and convection, which require direct or indirect contact, radiation is the transfer of energy in the form of electromagnetic waves. All objects that have a non-zero temperature emit infrared radiation in the form of photons. Hotter objects emit more photons than cooler ones. Through radiation, heat flows from the higher-temperature object to the lower one without the need for an intervening medium.

The sun warms the earth through the radiation of photons in the visible light spectrum. Heat radiators and heat lamps also function through infrared radiation. The rate of heat transfer by radiation is determined by factors like the temperature of the objects, their surface properties, and the distance between them. Darker, non-reflective surfaces are better at absorbing infrared radiation compared to light-colored, reflective surfaces, which absorb less heat through radiation.

While the three mechanisms of heat transfer exist independently in different situations, in practice, there is often a combination of processes at work simultaneously. For example, in a building, the outer walls transfer heat by both conduction through the solid material as well as radiation and convection of air trapped between internal and external surfaces. An important goal in building design is to minimize unwanted heat transfer and maximize desired heat flows using insulation, glazing, ventilation, and HVAC systems.

Now, having covered the major heat transfer mechanisms, we can examine specific HVAC components in more detail. For example, the role of insulation in addressing all three modes. The importance of air velocity and throw in delivering convection currents efficiently. How heat pumps leverage conduction, convection, and radiation together in one remarkable machine. By understanding fundamental heat transfer principles, HVAC technicians can diagnose and solve a wide range of problems faced in real systems. This knowledge forms a solid foundation for understanding modern HVAC equipment and designing effective, energy-efficient solutions.

Fluid Dynamics

Fluid dynamics plays an important role in many HVAC systems and components. Whether it's air moving through ductwork and registers or water circulating in pipes, understanding how fluids behave under different conditions is essential. In this section, we will explore some key principles of fluid dynamics and their applications in heating, ventilation, and air conditioning.

Key Principles

All fluids, whether gases or liquids, are made up of molecules that are always in constant motion. The study of fluid dynamics examines how these molecules interact and how the entire fluid mass flows and distributes pressure. One of the most fundamental principles is that molecules within a fluid will always move from areas of higher pressure to lower pressure until the pressures equalize.

Fluids also take on the shape of their container, meaning they have no fixed volume or shape of their own. This allows them to easily take any shape but also means even small variations in pressure or temperature can cause fluid flow. HVAC systems leverage positive and negative pressures to purposefully move air and water where needed. Whether it's a centrifugal fan blowing air through ducts or a boiler generating hot water pressure, understanding the pressures at play is key.

Another important characteristic of fluids is viscosity, which is a measure of their internal resistance to flow. Thicker liquids like motor oil are highly viscous, while thinner liquids and gases are less viscous and flow more freely. High-viscosity fluids require more energy to push or pump through pipes and tubes. Engineers must account for different fluid viscosities, selecting appropriately sized pumps and considering pressure drops that occur along piping runs or ductwork.

Airflow

Closely related to viscosity is laminar and turbulent flow. Laminar flow occurs when fluid particles move in parallel, smooth layers without disruption, like water flowing calmly in a garden hose. Such laminar motion results from low fluid velocities and viscosities. However, above a certain speed threshold, the smooth layers begin to interact irregularly, transitioning to turbulent flow. Turbulence greatly increases frictional resistance to flow, requiring more energy input. HVAC duct sizing seeks to avoid turbulence for efficient air conveyance.

One way fluids exchange momentum is through viscosity and friction along pipe or duct walls called the boundary layer. Molecules in direct contact are slowed by friction from the solid surface, transferring their slowed motion inward through continual molecular collisions. This effectively forms resisting laminar sub-layers next to all surfaces, with top layers slipping freely. In HVAC, smoothing interior surfaces and avoiding obstacles reduces form drag and boundary layer thickness for lower friction losses.

Fluid flow characteristics are described using specific terms and derived equations. The volumetric flow rate refers to cubic feet or gallons of fluid passing through a cross-section per minute. Volume and mass flow rates allow quantifying amounts moved. Velocity is the speed and direction that fluid particles travel, governed by pressure and viscosity. HVAC airflow delivery is often specified in cubic feet per minute (CFM) at certain velocities.

Perhaps the most important concept for HVAC design is Bernoulli's principle, which essentially states that as the velocity of a moving fluid increases, its pressure decreases. This is due to the conservation of energy as kinetic energy increases at the expense of pressure. Centrifugal fans create suction by precisely accelerating air into a lower-pressure region, while condenser fans do the opposite - gently pushing high-pressure air over coils. Pipes also rely on this process to convey both gas and liquid using pressure differentials.

Hydronic Systems

With a solid grounding in fluid dynamics basics, HVAC professionals can accurately predict system performance for air and hydronic distribution. Duct sizing software runs computational fluid dynamics (CFD) simulations, evaluating velocities, pressures, and temperature profiles throughout complex ductwork configurations. Likewise, piping networks are analyzed for pressure drops and proper component sequencing. Fluid dynamics also enables innovative new HVAC technologies such as thermosiphon cooling and low-pressure air distribution.

Applications

As new sustainable building standards drive tighter construction, fluid dynamics becomes even more important. Minimizing uncontrolled air leakage through duct pressure testing helps lower fan energy costs. Pipe insulation reduces conduction losses by blocking fluid heat conduction to surroundings. Variable speed pumps have onboard controllers responding to dynamic pressure needs. And energy recovery ventilators transfer up to 80% of heat between incoming and exhaust air streams using fluidized desiccant materials or heat exchangers.

Emerging Trends

With population growth urbanizing, massive central plants are emerging to serve the district heating and cooling needs of entire cities. Careful fluid dynamic modeling predicts the performance of miles-long piping distribution networks. Some bold new schemes even propose recovering waste heat from sewers or transportation tunnels using flowing water. Where conventional HVAC focuses on single buildings or campuses, emerging fluid dynamic systems optimize energy usage at regional scales.

CHAPTER 3
Types of HVAC Systems

Heating, ventilation, and air conditioning (HVAC) systems are fundamental components in modern buildings, ensuring occupants' comfort and environmental sustainability. In this chapter, we delve into the intricate world of HVAC systems, exploring the various types that play pivotal roles in regulating indoor climate. Each type holds unique features and functionalities, catering to diverse requirements and environmental conditions.

Heating Systems

Whether heating a single-family home or a large commercial building, selecting the right HVAC system is crucial to achieving comfort, efficiency, and indoor air quality goals. Heating accounts for over half of residential energy usage in cold climates, making system choice highly impactful. Let's explore some common options and factors influencing the decision.

Common Residential Systems

Natural gas furnaces remain ubiquitous due to low installation costs and relying on inexpensive, abundant fuel. Newer high-efficiency condensing models approaching 95% efficiency, utilizing a second heat exchanger to extract additional warmth from flue gasses. Proper venting directs exhaust safely outdoors while allowing combustion gases to pass through living areas presents safety risks. However, gas pipe infrastructure reaches many neighborhoods, and upgrading provides a cost-effective way to lower winter bills.

Modernizing with an air-source heat pump swaps combustion for evaporation, using refrigerants and vapor compression cooling technology now quite familiar to air conditioners. While relying on renewable electric power, heat pumps excel in moderate climates, extracting usable heat even from winter's chilliest days to provide both heating and cooling from one unit. The upfront costs compare favorably to a new furnace, thanks to incentives. Coupled with zoned delivery-like radiant flooring offering advanced comfort.

For all-electric homes, alternative electric technologies include high-efficiency ducted or ductless mini-split heat pumps, as well as simple resistance baseboard heating available virtually anywhere. Zoned electric systems deliver precise comfort control room by room. And heat pump hot water heaters offer dual duties of space and water heating for ultimate system synergy. While less eco-friendly than heat pumps, electric backup heat remains handy as a reliable supplement on the coldest nights or in emergency power outages unaffected by gas infrastructure issues.

Commercial and Industrial Systems

Hydronic systems circulating hot water prove a top choice for large commercial buildings due to high capacity and zoning flexibility. Boilers heat water distributed through insulated piping to convectors, radiators, or radiant floor/ceiling panels, blending efficiency with structural heating. Modular zoning allows individual temperature settings. Integrating with solar thermal or seasonal underground thermal energy storage (UTES) takes advantage of renewable heat sources. Larger facilities often centrally producing both hot water and steam for maximum energy recapture benefit from economies of scale.

Heat recovery technologies recover waste heat to boost efficiency. Exhaust air heat pumps (ERVs) transfer warmth between incoming fresh and outgoing exhaust ventilation streams. Runaround coils in constant circulation isolate waste heat to reclaim as much as 80% from vented air or mechanical equipment before being released outdoors. Thermal wheels within air handlers' absorption and release heat more directly between air streams. Reclaiming exhaust heat from cooking equipment, luminaries, people, or processes provides free "bonus" heat to condition spaces without adding new fuels.

Advanced distributed systems offer cutting-edge solutions. Geothermal or water-source heat pumps tap into the earth's natural heating/cooling properties through buried ground loops or surface water heat exchangers. District heating/cooling networks on an urban scale allow centralized cogeneration plants to produce both electricity and thermal energy delivered via an underground network of insulated pipes to numerous buildings. This centralizes pollution control while maximizing fuel efficiency.

Selecting the best HVAC heating system involves weighing efficiency, capital costs, operating expenses, indoor air quality, appropriate capacity, and comfort control capabilities. Many factors influence these metrics, from climate characteristics and fuel availability to building design and zoning approach. With so many excellent system choices available today, open discussions help determine the optimal solution balancing investment, performance, and sustainability for each unique project. Proper installation verifies safe, reliable operation, protecting occupants for many heating seasons to come.

Cooling Systems

As summer temperatures rise across many regions, reliable cooling remains essential for both residential homes and large commercial buildings. Luckily, significant advancements in HVAC technology now provide high-performance, energy-efficient, and sustainable air conditioning options. Let's explore some of the most prevalent cooling system choices in today's market.

Central air conditioners using refrigerant vapor compression cycles remain popular for their ubiquity and simple designs. However, new super-efficient variable speed units push past standard SEER ratings. Inverter-driven compressors precisely regulate output according to real-time cooling loads for maximum comfort with minimal energy waste. Thicker insulation in unit cabinets and updated refrigerants increase already high coefficient of performance (COP) values.

Proper sizing and ductwork design prove crucial, as oversized equipment short-cycles inefficiently while undersized units cannot keep up with demands. Computerized manuals assist professionals with accurate load calculations considering factors like envelope insulation, window orientation, number of occupants, and local climate data. Duct sealing compounds and mastic patches further tighten distribution systems, preventing leakage of up to 30% cooling capacity.

Smart thermostats take efficiency another step via internet connectivity. Mobile apps allow remote temperature adjustments from anywhere. Sensors track occupancy, outdoor conditions, and unit run-times to learn optimal schedules tailored for each home. Algorithms minimize cycles through smart setpoints, deactivating systems only as needed for maximum savings of up to 23% annually. Voice commands integrate with smart home automation for touchless convenience.

Versatile mini-split heat pump systems provide both heating and cooling while allowing flexible zoning control. Their smaller variable-speed inverter compressors deliver heating down to -13°F and cooling on the hottest days. New multi-head outdoor units consolidate piping for simplified installation, serving multiple indoor wall-mounted cassettes independently per room. An optional heat recovery ventilator maintains superior indoor air quality, exchanging heat and moisture between ventilation air streams up to 93%.

Larger buildings rely on central water chilling systems using electric or gas-powered chillers. Magnetic bearing centrifugal chillers eliminate friction for greater efficiencies, exceeding prior industrial standards by 50%. Modular additions permit redundancy, while sophisticated building automation optimizes operations according to occupancy-based schedules. Thermal energy storage tanks catch overnight chill for gradual release, meeting daytime peak loads. In all, such commercial technologies cut energy usage by up to 30%.

Geothermal heat pump systems tap underground temperatures via buried ground loop fields or surface water exchangers for the highest efficiency cooling available. A single compact outdoor unit provides four-season climate control for the entire home without fossil fuel emissions. They serve larger structures as well by exchanging heat within centralized water loops. Though installation costs are higher initially, ultra-low operating expenses pay back the investment within 7-10 years on average in utility savings alone.

Advanced techniques like radiant cooling panels provide customized comfort, removing heat via radiation rather than airflow. Critically, they eliminate drafts and MIXED issues inherent to conventional fans while achieving equal performance at higher temperatures. Combining radiant systems with IGBT variable refrigerant flow splits adapts HVAC flexibly throughout repurposed older buildings. Control retrofits integrate legacy zoning without invasive renovations.

Cutting-edge technologies decentralize utilities, pushing renewable efficiency. Solar thermal collectors harvest daylighting to precondition ventilation intake or chill secondary loop fluid, dropping condenser and chiller loads by 50%. Integrated photovoltaics now directly power mini-split inverters from rooftop arrays. And combined heat/power microgrids ensure resilience via cogeneration fuel cells serving neighborhoods entirely with clean district energy.

Modern HVAC choices incorporate high performance, occupant wellness, sustainable design, and smart analytics altogether through customizable solutions. Selecting the optimal, tailored system involves weighing needs unique to each application, from climate to structure to lifestyle considerations. Advanced options now affordably condition spaces through efficient flexibility and renewable resources available anywhere.

Ventilation Systems

Heating, ventilation, and air conditioning (HVAC) systems play a vital role in commercial buildings by maintaining thermal comfort and indoor air quality. While heating and cooling capabilities are commonly known attributes of HVAC, proper ventilation is perhaps the most critical function for occupant health, safety, and productivity. Ensuring adequate ventilation involves distributing and replacing indoor air through ventilation systems, which can range from simple solutions to complex engineered designs depending on building size and occupancy needs.

Why Is HVAC Ventilation Important?

The main purpose of ventilation is to provide occupants with fresh outdoor air while removing indoor contaminants like odors, dust, chemicals, and potentially harmful gases or biological agents like viruses or bacteria. Too little ventilation can compromise indoor air quality and lead to higher concentrations of these pollutants. This poses health risks like sick building syndrome with symptoms like headaches, fatigue, and difficulty concentrating. Poor ventilation also diminishes comfort and makes buildings prone to stale, stuffy air.

Modern commercial buildings have fewer opportunities for natural ventilation through open windows or doors compared to older structures. This makes mechanical ventilation through HVAC systems essential for maintaining acceptable indoor air quality. Proper ventilation helps dilute airborne pathogens, control humidity levels, and create a more appealing indoor environment where people can be productive. It also prevents harmful indoor pollutants from accumulating to hazardous levels over extended periods of occupancy.

Ventilation is especially critical now with the COVID-19 pandemic highlighting the importance of exchanging indoor air. Recent studies indicate the virus can remain suspended in air and accumulate indoors, posing infection risks if air is recirculated without adequate filtration or dilution through fresh air intake. Well-designed HVAC systems equipped to deliver the proper amount of outside air replacement are fundamental for mitigating airborne disease transmission in public and commercial spaces.

Key Components of Commercial HVAC Ventilation Systems

Most commercial HVAC systems incorporate dedicated ventilation components to ensure adequate fresh air delivery. Components vary depending on design complexity but commonly include:

- Air Handling Units: Large central equipment that conditions circulates, and distributes air throughout a building via ductwork. AHUs control ventilation by mixing fresh outdoor air with return air prior to temperature and humidity adjustment.

- Supply and Return Air Grilles/Registers: Wall-mounted or ceiling-mounted grilles that discharge conditioned air into spaces (supply) and pull air back to AHUs for reconditioning (return). Proper placement is critical for effective air distribution and contaminant removal.
- Ductwork: An underground maze of insulated metal or fiber ducts that moves supply and return air between AHUs and individual grilles/registers throughout the building envelope.
- Fresh Air Intakes: Outside openings with louvers, dampers, and air filters that pull a preset amount of outdoor air into AHUs for ventilation and conditioning.
- Thermostats: Wall-mounted sensors that detect space temperature and humidity for AHU control. Some advanced versions also monitor CO2 or VOC levels.
- Ventilation Controls/Building Automation Systems: Computer programs that regulate AHU ventilation rates based on occupancy, time of day, outdoor conditions, indoor air quality sensor readings, and other customizable settings.

Ventilation Design Considerations

When designing HVAC ventilation for commercial buildings, engineers must carefully assess the anticipated internal loads and occupant densities to size systems properly. Key factors addressed include:

- Occupancy/Space Types: Ventilation needs vary greatly between office, retail, restaurant, and other occupancy types. Occupant schedules are also considered.
- Outdoor Air Requirements: Code-mandated minimum fresh air intake volumes based on building/space use are established (e.g., 15 cubic feet per minute (CFM) per person in offices).
- Indoor Contaminant Sources: Processes, furnishings, and chemicals that may off-gas require more dilution through higher ventilation levels.
- Energy/Operational Costs: Over-ventilating wastes energy while under-ventilating compromises health. Optimal rates balance both.
- Filtration/Air Cleaning: MERV-rated filters or other cleaning technologies address indoor/outdoor pollutants.
- Zoning/Thermal Comfort: Dividing buildings into climate-controlled zones for pressure/temperature control enhances comfort.
- Controls/Monitoring: Automated control sequences optimize ventilation efficiency through sensors and building automation integration.
- Installation/Distribution: Proper equipment sizing, duct sizing/layout, and diffuser placement ensure air reaches all areas effectively.

By accounting for these factors, HVAC engineers design tailored ventilation solutions customized for each unique building and its operating profile. Their goal is to establish ventilation strategies that satisfy indoor environmental quality expectations as cost-effectively as possible.

Common Ventilation System Types

There are different commercial HVAC ventilation system approaches depending on the building scale:

- Single Zone: Smaller buildings are typically served by a single central air handler providing uniform conditioning to the entire interior space. Ventilation is introduced via the AHU fresh air intake.
- Variable Air Volume (VAV): Larger buildings implement VAV, where the AHU supplies air to variable dampers controlling zone-level flow. This enhances zoning and efficiency.
- Dedicated Outdoor Air Systems (DOAS): In very large commercial facilities, a separate DOAS dedicated solely to ventilation preconditions outdoor air independent of thermal conditioning needs. This isolates and optimizes ventilation management.
- Heat Recovery Ventilators (HRV)/Energy Recovery Ventilators (ERV): For both new construction and retrofit applications, HRVs/ERVs employ enthalpy exchange cores to precondition ventilation air and recapture heating/cooling capacity through heat/energy transfer between extracted and supply air streams. This boosts efficiency in moderate to cold climate zones.

Proper Maintenance Ensures Ongoing Performance

Regardless of the specific system design, diligent HVAC maintenance is crucial for ventilation systems to operate as intended over the long run. Some key maintenance best practices include:

- Filter Replacement: Clogged filters hinder designed airflow and reduce the filter's ability to remove contaminants. Change filters per manufacturer schedule.
- Drain Pans/Piping: Ensure condensate can drain properly to prevent mold/microbial growth that degrades indoor air. Check for leaks.
- Coil Cleaning: Dirty evaporator and condenser coils lose heat transfer ability and energy efficiency. Professionally clean annually.
- Belt/Component Inspections: Check pulleys, motors, dampers, and linkages are moving freely with no obstructions quarterly. Lubricate or replace worn parts.
- Calibration/Tune-Ups: HVAC technicians perform seasonal check-ups and calibration of sensors, controls, and ventilation settings/schedules to maintain IAQ performance parameters.
- Record Keeping: Documenting maintenance, repairs, and filter changes establishes operational history useful for troubleshooting or warranty claims down the road.

When properly designed, installed, and maintained, commercial building HVAC ventilation systems effectively deliver the fresh air needed to support occupant wellness, productivity, and indoor environmental quality long into the future. Though often overlooked, ventilation is truly the lifeblood keeping buildings healthy.

Hybrid Systems

For decades, HVAC design essentially involved choosing between unitary equipment like air conditioners, heat pumps, boilers, or chillers to fit a building's needs. However, a growing recognition that no single technology optimally serves every application is driving innovation toward "hybrid" HVAC systems. By thoughtfully combining different HVAC components, hybrid systems leverage each element's strengths while overcoming individual limitations.

Early adopters report significant performance gains from custom-engineered hybrid solutions. As the paradigm evolves, hybrid HVAC shows promise to revolutionize sustainable and cost-effective conditioning. This article explores the rise of hybridization, common system combinations demonstrating success, important design considerations, and an exciting future as the approach matures.

The Concept of HVAC Hybridization

At their core, HVAC hybrid systems strategically pair two or more proven HVAC equipment categories. Whereas unitary systems may falter in certain conditions, a well-designed hybrid aims to derive maximum efficiency across all load and climate scenarios.

For example, an air source heat pump functions best between 35-75°F but loses efficiency outside that range. Pairing it with a gas furnace minimizes expensive electric resistance heat operation below 35°F while keeping equipment sized appropriately. During milder winter weather, the heat pump handles most of the load efficiently.

Other common hybrid concepts include combining boilers or fossil fuel generators with renewable power like solar, pairing chillers or heat pumps with geothermal/surface water, or incorporating energy recovery ventilators alongside chillers. All serve to optimize performance using targeted synergies.

Hybrids also further energy goals by facilitating load shifting. For instance, if a building generates solar power during sunny midday periods, an ice-storage hybrid chiller can capitalize on that surplus renewable energy by producing chilled water overnight for storage. That stored "ice" then cools the building during peak daytime electric rates, easing the strain on the grid.

Realizing Significant Energy and Cost Savings

Early industry case studies provide promising proof points that properly designed hybrid systems can boost building efficiency by 15-30% or more over conventional unitary alternatives. Some revealed projects:

A hospital in Canada paired its boiler plant with a seasonal thermal energy storage (STES) system and 176-ton absorption chiller for 60% cheaper chilled water production versus electric reciprocating chillers alone. The hydronic hybrid also better leveraged the facility's waste heat streams.

At a New York university, switching from electric heat pumps and boilers to a hybrid gas heat/geothermal VRF solution reduced heating costs by 20% and lowered cooling expenses by 5% versus an all-electric baseline. Carbon emissions decreased by 30% as well.

An office park in California deployed a heat recovery chiller alongside a gas-fired absorption chiller and heat pumps. Compared to standard systems, the integrated hybrid slashed total energy costs by an estimated 35% while maintaining occupant comfort reliably.

Key Design Considerations for Success

Thoughtful engineering lies at the heart of a high-functioning HVAC hybrid system. Core factors evaluated include:

- Building Construction/Envelope: A tightly-sealed, well-insulated building minimizes unnecessary additional load on HVAC equipment.
- Occupancy Schedules: Understanding occupancy patterns helps optimize operations based on actual versus theoretical loads.

- Outdoor Climate Conditions: Geographic weather extremes dictate the best supporting/supplementary hybrid technologies.
- Indoor Load Profiles: Process loads, IT equipment, plug loads, lighting, and ventilation needs factor into sizing.
- Available Energy Sources: Options like gas, electric, geothermal, solar power, sewage/waste heat feed selection.
- Existing Infrastructure: Retrofitting may modify versus replace, while new builds offer complete design freedom.
- Life-Cycle Cost Analysis: Weighing upfront costs versus long-term energy/maintenance expenses ensures optimized value.
- Control Strategies: Sophisticated logic optimally coordinates hybrid components through varied conditions.

With a thorough understanding of these elements, engineers craft adept hybrid configurations, raising resilience, lowering operating budgets, and advancing sustainability targets for each unique project. Advanced building modeling aids optimization.

Evolving System Combinations and Integration

As the hybrid HVAC concept matures, additional innovative combinations continue emerging to fit wider applications:

- Fuel Cells Paired with Heat Pumps: Fuel cells' waste heat boosts heat pump efficiency for colder climates in a decentralized microgrid package.
- CO_2 Heat Pumps and Lithium-Ion Battery Storage: Leveraging thermal energy storage for COP improvements captures unused renewables cheaply.
- VRF Integrated with Ground/Water Source Heat Pumps: Geoexchange field and proprietary inverters synergize for peak performance.
- Water-Cooled Chillers and Cooling Towers Hybridized: The Cooling tower acts as a thermal battery, pre-chilling water at off-peak times.
- Desiccant Dehumidification Hybrid with Heat Recovery: Recovers "free" cooling and dehumidification in tandem from exhaust air streams.

Further, researchers explore subsystem interactions like combining evaporative pre-cooling with economizers or integrating dedicated outdoor air systems (DOAS) alongside active chilled beams. Smarter controls also smooth optimized coordination.

As these novel hybrids prove themselves, modular packaged solutions tailored for diverse building stock will streamline design, installation, and commissioning further. The expanding knowledge base primes HVAC for its next transformative technological stage.

An Optimistic Future for Sustainability

If custom-engineered hybrid systems already deliver such benefits in their infancy, the future potential appears bright as the field matures. Experts envision hybrid HVAC:

- Boosting renewables adoption by stabilizing intermittent solar and wind with storage, generators, and two-way power capabilities.
- Facilitating electrification through hybrid heat pumps is able to selectively backup electric heating/cooling systems versus traditional fuels during peaks.

- Optimizing via machine learning and AI coordinating complex hybrid organizations autonomously based on massive operational datasets.
- Scaling deployment through standardized hybrid modules and components permits simpler retrofits of older equipment at end-of-life.
- Gaining mainstream consideration for new construction to maximize paybacks upfront through performance-based design evolution.

As hybrid HVAC matures, aggressively reducing environmental impacts while improving affordability, reliability, resilience, and comfort suggests this avenue is primed to revolutionize sustainable building systems globally. With open-minded support, the hybrid frontier is just getting started.

PART II
RESIDENTIAL HVAC

CHAPTER 4
Components and Design

Thermostats

Thermostats, small yet potent devices, stand as the nerve center of any heating, ventilation, and air conditioning (HVAC) system. These unassuming components wield immense power, dictating the comfort levels within a space by skillfully managing temperature settings. A well-designed thermostat, in essence, is the guardian of our indoor environment, entrusted with the crucial task of maintaining a harmonious balance of warmth and coolness.

Types of Thermostats

In the vast world of thermostats, diversity reigns supreme. The technological evolution has birthed several distinct types, each tailored to meet specific needs.

Mechanical Thermostats

These are the trailblazers of the thermostat realm. Mechanical thermostats, with their rudimentary yet reliable design, employ a simple dial or slider mechanism. Their core utilizes a bimetallic coil that bends and straightens in response to temperature changes. The bending of this coil triggers the activation of the heating or cooling system, maintaining the desired temperature.

Digital Thermostats

The advent of digital technology brought forth a new era in thermostat evolution. Digital thermostats, equipped with a digital display and intuitive buttons, revolutionized temperature control. They allow for precise temperature settings and often feature programmable capabilities. With programmable settings, users can orchestrate temperature schedules for different times of the day, tailoring comfort according to their daily routines.

Smart Thermostats

Smart thermostats, the epitome of modern innovation, have taken the art of temperature control to unprecedented heights. These intelligent devices seamlessly integrate into our interconnected world. They can be accessed and controlled remotely through smartphone applications or via voice commands through smart home systems like Amazon Alexa or Google Assistant. Smart thermostats are not merely reactionary; they possess adaptive algorithms that learn our preferences and habits, optimizing energy consumption and providing valuable energy-saving suggestions.

Working Mechanism

The fundamental principle governing thermostats is elegantly simple yet highly effective. At its core lies a temperature sensor, diligently monitoring the current temperature of the environment. This real-time temperature data is then compared to the predetermined set temperature. When the current temperature deviates from this set threshold, the thermostat sends a signal to the heating or cooling system, prompting it to activate or deactivate, ultimately bringing the environment back to the desired temperature.

Importance in HVAC Systems

Thermostats serve as the custodians of energy efficiency within a space. By meticulously regulating heating and cooling systems, they aid in minimizing energy consumption, leading to reduced utility costs. Moreover, the capabilities of modern smart thermostats to learn user preferences and optimize heating and cooling cycles further bolster energy efficiency. This fusion of technological prowess and environmental consciousness makes thermostats not just controllers of comfort but also champions of sustainability.

Ductwork

Ductwork is the intricate network of channels or pipes that form a crucial part of heating, ventilation, and air conditioning (HVAC) systems. It serves as the circulatory system of the HVAC setup, enabling the efficient distribution of conditioned air throughout a building or space. These ducts ensure that heated or cooled air reaches its intended destination in a controlled and consistent manner.

Types of Ductworks

Ductwork comes in various types, each designed for specific applications and tailored to suit different environments. Understanding the types of ductwork is essential for selecting the most suitable option for a particular HVAC system. Here are the primary types of ductwork:

Sheet Metal Ducts

Sheet metal ducts are among the most common and widely used types of ductwork. Typically constructed from galvanized steel sheets, they offer durability and strength. Their rigidity allows for extended use without deformation under pressure. Sheet metal ducts are well-suited for both residential and commercial applications, making them a versatile choice in HVAC installations.

Fiberglass Ducts

Fiberglass ducts, constructed from fiberglass-reinforced panels, are a lightweight and economical alternative to sheet metal ducts. They are lined with fiberglass insulation, providing excellent thermal and acoustic properties. The insulation not only enhances energy efficiency but also reduces noise transmission through the ducts. These ducts are relatively easy to install and handle, making them a preferred choice in a variety of HVAC installations.

Flexible Ducts

Flexible ducts, as the name implies, are flexible and versatile in their applications. They are often utilized in areas where rigid ducts are impractical or difficult to install. The flexibility of these ducts allows for easy routing in constrained or irregular spaces. However, it's essential to ensure proper support and avoid sharp bends during installation to maintain optimal airflow and efficiency.

Duct Board

Duct board is made from fiberglass boards and is a lightweight alternative to sheet metal ducts. These boards are easy to cut, shape, and assemble, facilitating quick and efficient installation. Duct boards provide good insulation and reduce the risk of air leakage, contributing to energy efficiency. They are particularly effective when installed in areas with limited space or where rigid ducts are challenging to use.

Design Considerations

Designing an efficient and effective ductwork system is a complex process that involves careful consideration of several factors. The design should ensure optimal airflow, energy efficiency, and even distribution of conditioned air. Here are the key design considerations for ductwork:

Duct Size and Shape

Determining the appropriate size and shape of ducts is crucial to achieving the required airflow for each space. Ducts that are too small may lead to restricted airflow, affecting the system's efficiency, while oversized ducts can cause pressure imbalances and increased energy consumption.

Insulation

Proper insulation of ductwork is vital to prevent heat gain or loss as air travels through the system. Insulation helps maintain the desired temperature of the conditioned air, enhancing overall energy efficiency and reducing heating and cooling costs.

Airflow Balancing

Achieving a balanced airflow throughout the building is a key consideration in ductwork design. This involves adjusting damper settings and duct sizes to ensure consistent air distribution maintaining uniform temperatures in all rooms. Properly balanced airflow contributes to optimal system performance and comfort.

Duct Material Selection

Choosing the appropriate material for the ducts is essential for durability, efficiency, and compatibility with the HVAC system. Factors such as cost, the type of building, and the intended application influence the material selection.

Sealing and Air Leakage Prevention

Proper sealing of joints and seams is crucial to prevent air leakage in the ductwork. Air leakage can significantly reduce system efficiency and lead to energy wastage. Employing sealing techniques and using quality sealants during installation helps maintain airtight ducts.

Maintenance and Cleaning

Regular maintenance and cleaning of ductwork are fundamental to ensuring optimal performance and preserving indoor air quality. Over time, dust, debris, mold, and other contaminants can accumulate within the ducts, hindering airflow and potentially posing health risks. Here are the key aspects of maintenance and cleaning:

Scheduled Inspections

Regular inspections of the ductwork are essential to identify any signs of damage, wear, or deterioration. Professionals can assess the condition of the ducts, insulation, and sealing to determine if repairs or maintenance are required.

Cleaning

Periodic cleaning of ducts is crucial to remove accumulated dust, allergens, and debris. Professional duct cleaning services use specialized equipment to thoroughly clean the ductwork, ensuring improved indoor air quality and system efficiency.

Filter Replacement

Timely replacement of air filters is a vital aspect of ductwork maintenance. Clogged or dirty filters obstruct airflow, forcing the HVAC system to work harder and consume more energy. Regular filter replacement promotes optimal system performance and extends the life of the HVAC components.

Sealing Repairs

Any signs of leaks or damaged seals should be promptly addressed to maintain the integrity of the duct system. Repairs may involve resealing joints or replacing damaged sections to prevent air leakage and maintain energy efficiency.

Filters and Air Quality

Filters and their role in maintaining indoor air quality are often underestimated in HVAC systems, yet they are indispensable components that play a crucial role in ensuring not only a comfortable living or working environment but also the health and well-being of the occupants.

Types of Filters

Filters come in various types, each designed to address specific needs in terms of particle capture and air quality enhancement. Understanding the distinctions between these filter types is essential for selecting the most suitable one for a particular HVAC system and the environment it serves.

Fiberglass Filters

These filters are perhaps the most common and cost-effective choice for many HVAC systems. Typically constructed with layered fiberglass strands, these filters are designed to capture larger particles such as dust, lint, and debris effectively. However, they are less efficient at trapping smaller particles and allergens. Fiberglass filters are disposable and should be replaced regularly.

Pleated Filters

Pleated filters, often made from cotton or polyester, are superior to fiberglass filters when it comes to filtration efficiency. Their accordion-like design provides a larger surface area for trapping particles. This means they can capture smaller particles, including pet dander, pollen, and mold spores, making them a better choice for improved air quality.

HEPA (High-Efficiency Particulate Air) Filters

HEPA filters are the gold standard in air filtration. These filters meet stringent standards set by the U.S. Department of Energy, capable of capturing 99.97% of particles as small as 0.3 microns in size. This includes bacteria, viruses, and fine dust. HEPA filters are often used in environments where air quality is of utmost importance, such as hospitals, laboratories, and cleanrooms.

Electrostatic Filters

Electrostatic filters use an electric charge to attract and capture particles. These filters can be washable or disposable, depending on the design. They are effective at capturing a wide range of particles, including allergens, and can be a more eco-friendly option as they can be reused.

Activated Carbon Filters

These filters are specifically designed to remove odors and gases from the air. They contain a layer of activated carbon, which absorbs volatile organic compounds (VOCs), unpleasant odors, and even some chemicals. Activated carbon filters are commonly used in households to improve indoor air quality.

UV-C Filters

Ultraviolet-C (UV-C) filters use ultraviolet light to kill bacteria, viruses, and mold spores that pass through the filter. While they are not typically used as stand-alone filters, they can be integrated into HVAC systems to provide additional air purification.

Importance of Filters

The significance of filters in HVAC systems extends far beyond just trapping particles. Their role in enhancing indoor air quality is multifaceted and crucial for several reasons:

Particle Removal

Filters act as the first line of defense against airborne particles. They capture dust, pollen, pet dander, and other allergens, preventing them from circulating in the indoor air. This is especially important for individuals with allergies or respiratory conditions.

Health Benefits

Improved air quality, achieved through the efficient filtration of particles and contaminants, has direct health benefits. It reduces the risk of respiratory problems, allergies, and other health issues related to poor indoor air quality.

Protection of HVAC System

Filters also play a significant role in safeguarding the HVAC system itself. By preventing particles from entering sensitive components like coils and fans, filters help maintain the system's efficiency and longevity.

Odor and Gas Removal

Filters such as activated carbon filters are effective at removing unpleasant odors and gases, contributing to a more pleasant and healthier indoor environment.

Energy Efficiency

Clean filters allow for better airflow through the HVAC system, reducing the workload on the system's fan and blower. This, in turn, enhances energy efficiency and can lead to lower utility bills.

Environmental Impact

In the case of washable or reusable filters, there is an environmental benefit as well. These filters reduce the need for disposable filters, lowering waste and contributing to a more sustainable approach to HVAC maintenance.

Filter Replacement and Maintenance

Proper maintenance and regular replacement of filters are essential to ensure their optimal performance and to reap the benefits mentioned above. Neglecting filter maintenance can lead to a range of issues, including decreased indoor air quality, reduced energy efficiency, and potential damage to the HVAC system. Here are some key aspects of filter maintenance:

Replacement Frequency

The frequency of filter replacement depends on several factors, including the type of filter, the environment, and the manufacturer's recommendations. In general, disposable filters should be replaced every one to three months. However, it's advisable to check the filter monthly, especially during heavy-use periods, and replace it if it appears dirty or clogged.

Check for the Right Size

When replacing filters, it's crucial to ensure you are using the correct filter size that matches the dimensions specified by the HVAC system's manufacturer. Ill-fitting filters can allow unfiltered air to bypass the filter, diminishing its effectiveness.

Cleaning (for Washable Filters)

If your HVAC system uses washable filters, they should be cleaned regularly according to the manufacturer's instructions. This typically involves rinsing or vacuuming off accumulated dust and debris. Ensure the filter is completely dry before reinserting it into the system.

Consider Upgrading

If you are concerned about air quality, consider upgrading to a higher-efficiency filter, such as a pleated filter or a HEPA filter, if your HVAC system supports it. However, be aware that high-efficiency filters may require more frequent replacement due to increased particle capture.

Professional Inspection

Periodic professional inspection and maintenance of the entire HVAC system, including the filter, are recommended. HVAC technicians can assess the condition of filters and provide guidance on maintenance practices to optimize system performance.

Monitor Air Quality

Invest in an air quality monitor for your indoor space. This device can provide real-time data on air quality, including particle levels, humidity, and VOC concentrations. It can help you assess the effectiveness of your HVAC system's filtration and guide your maintenance schedule.

CHAPTER 5
Installation and Maintenance

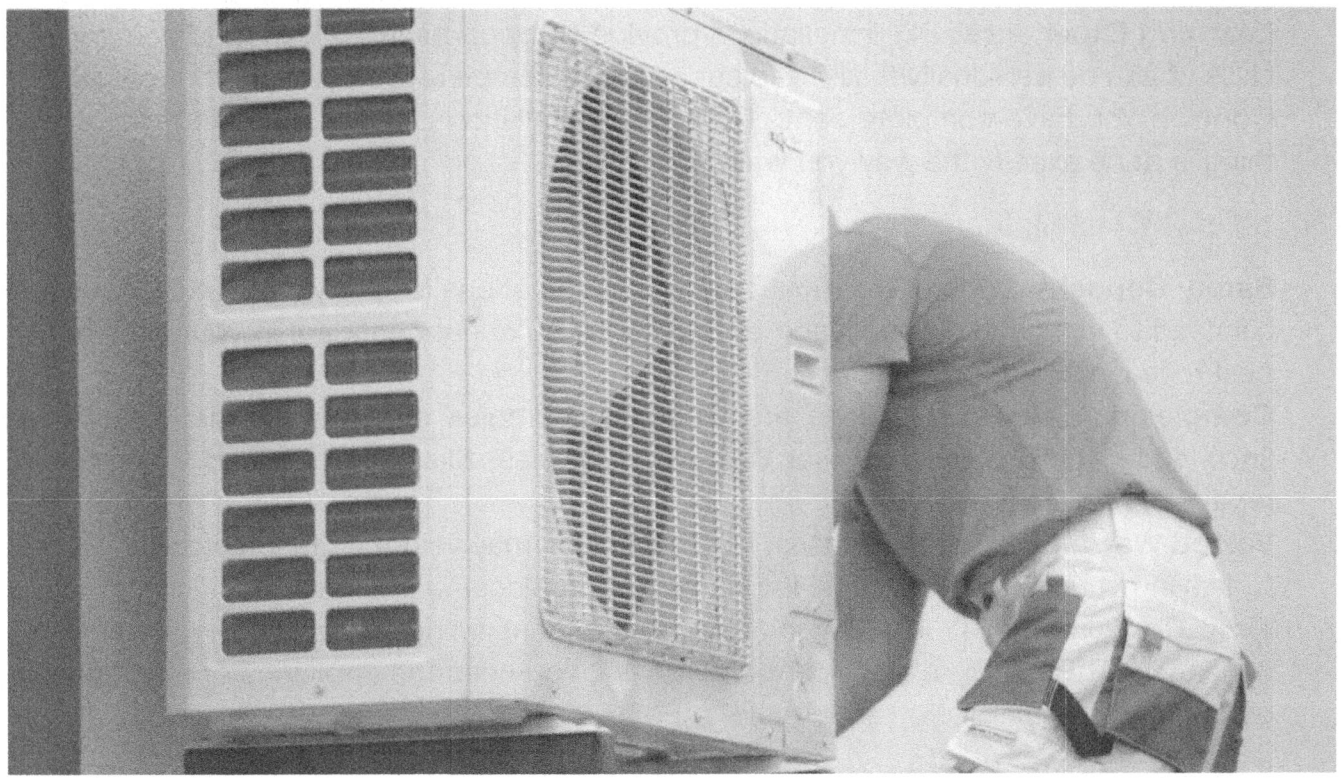

In the world of appliances and machinery, installation and maintenance are two critical aspects that can significantly impact the performance, longevity, and safety of your equipment. In this chapter, we will delve deep into the intricacies of installation and maintenance, discussing the choice between DIY (Do-It-Yourself) and professional installation, the importance of routine maintenance, and basic troubleshooting techniques. By the end of this comprehensive exploration, you will have a thorough understanding of how to ensure your appliances and machinery not only work optimally but also have a prolonged lifespan.

DIY vs. Professional Installation

The DIY Approach: Pros and Cons

When it comes to installing new appliances or machinery, many homeowners are tempted by the prospect of saving money by going the DIY route. While this approach can be rewarding and cost-effective for certain tasks, it's essential to consider the pros and cons carefully.

Pros of DIY Installation

- **Cost Savings:** The most apparent advantage of DIY installation is the potential for cost savings. You don't have to pay for professional labor, which can be a significant portion of the overall cost.
- **Flexibility:** DIY installations allow you to work on your schedule. You don't have to wait for a technician to become available, which can be especially beneficial in emergencies.
- **Learning Experience:** DIY installations provide an opportunity to learn and acquire new skills. It can be personally fulfilling to successfully complete a project on your own.
- **Control:** You have complete control over the installation process, ensuring that everything is done exactly the way you want it.

Cons of DIY Installation

- **Safety Concerns:** One of the most critical considerations is safety. Incorrect installation can lead to accidents, injuries, and even fires. If you're not confident in your abilities, it's best to leave it to the professionals.
- **Complexity:** Some installations are inherently complex and may require specialized knowledge, tools, or permits that you don't possess. Mistakes can lead to expensive repairs or replacements.
- **Voided Warranties:** In some cases, DIY installation may void the manufacturer's warranty, leaving you responsible for all future repair or replacement costs.
- **Time-Consuming:** DIY installations can be time-consuming, especially if you're not experienced. This can be a significant drawback if you need the appliance or machinery to be operational quickly.

The Professional Installation Option

Professional installation, on the other hand, involves hiring experts who specialize in installing specific types of appliances or machinery. Here are the key aspects to consider:

Pros of Professional Installation

- **Expertise:** Professionals are trained and experienced in their field. They know the ins and outs of installation, ensuring that everything is set up correctly and safely.
- **Safety:** Professional installers are well-versed in safety protocols and standards. This reduces the risk of accidents or mishaps during and after installation.
- **Efficiency:** Professionals work efficiently and can complete installations quickly, reducing downtime and inconvenience.
- **Warranty Preservation:** In most cases, professional installation maintains the manufacturer's warranty, providing peace of mind in case of defects or issues down the line.

Cons of Professional Installation

- **Cost:** Hiring professionals typically comes with a cost, which can be a significant factor for budget-conscious individuals.
- **Scheduling:** You may need to schedule the installation at a time that suits the technician, which might not align with your preferred timeline.

- **Less Control:** While professionals ensure a high-quality installation, you may have less control over the process compared to a DIY approach.

Making the Choice

The decision between DIY and professional installation ultimately depends on various factors, including the complexity of the installation, your skill level, budget, and time constraints.

For straightforward installations like assembling furniture or installing a new light fixture, DIY can be a practical and rewarding choice. However, for more complex tasks such as wiring electrical appliances, plumbing, or installing heavy machinery, it's often safer and more efficient to hire professionals.

When deciding, always consider your safety and the long-term implications of the installation. It's often better to invest in professional installation upfront to avoid costly repairs or accidents in the future.

Routine Maintenance

Regular maintenance is vital for any equipment, be it household appliances, industrial machinery, or HVAC systems. Performing routine upkeep has several key benefits that contribute to long-term performance and savings.

Firstly, consistent maintenance prolongs the lifespan of equipment. By addressing minor issues early on, you prevent small problems from escalating into major repairs down the line. Replacing filters, tightening loose parts, and cleaning dust buildup seem simple, but these small tasks add up to reduce wear and tear. Equipment that receives routine care can often operate for years beyond its expected lifespan compared to neglected machines.

Secondly, diligent maintenance keeps equipment working at peak efficiency. As an example, dirty condenser coils force air conditioners to work harder to expel heat, wasting energy and straining components. Regular cleaning maintains free airflow to boost efficiency. Industrial machinery also produces maximum throughput when well-oiled and calibrated. Efficiency gains generally translate into cost savings on energy and operation expenses.

Another crucial benefit is enhanced safety. Routinely inspecting for faulty wiring, leaky gas connections, worn parts, or other hazards reduces risk. Identifying and resolving potential dangers before they cause equipment failures prevents accidents and disasters. Safety must be the top priority with maintenance work itself as well.

While preventive maintenance requires an initial time investment, it pays long-term dividends in avoided repair costs. Prioritizing routine care leads to fewer unexpected breakdowns and expensive reactive servicing. Maintaining equipment maximizes savings and minimizes inconvenience.

To begin reaping these benefits, start by creating a maintenance schedule documenting the required tasks and intervals for each appliance or machine. Consult manufacturers' guidelines for specific recommendations. Be diligent in sticking to the schedule while tracking all maintenance activities for auditing. Safety should always come first during maintenance through powering down equipment, using protective gear, and following precautions.

Routine yet comprehensive maintenance keeps equipment humming along reliably, safely, efficiently, and economically for years of faithful service. Investing a little time up front prevents untold costs and headaches down the road.

Basic Troubleshooting

No HVAC system runs perfectly forever. Despite routine maintenance, occasional issues can arise that compromise climate comfort and efficiency. Having fundamental troubleshooting skills empowers homeowners to diagnose and resolve many basic HVAC problems without waiting for professional assistance. Let's explore tips for tackling common residential HVAC troubleshooting.

General Troubleshooting Approach

Approaching any malfunction systematically is key. Start by clearly defining the symptoms – is it a lack of heating or cooling, reduced airflow, unusual noises, or other anomalous behavior? Note when the issue began and any recent events like storms that might have contributed. Consult the owner's manual, which often describes common problems and solutions. Visually inspect accessible components like air filters and ductwork for obstruction.

If the issue isn't obvious, methodically isolate potential causes through testing. Turn the thermostat to different settings to check if the system responds properly. Feel air vents to verify adequate airflow. Listen for inconsistent noises that signal faulty components. Switch between heating and cooling modes to pinpoint the problem. Comparing normal and abnormal operations narrows down the culprit.

Online HVAC communities prove invaluable troubleshooting resources. Search forums and boards describing your specific model and symptoms. Obtaining advice from fellow homeowners who have encountered similar problems can reveal effective resolutions. However, take care to consult reputable sites and verify suggestions are safe for your system.

Common Troubleshooting Tips

While troubleshooting varies by HVAC system, some general tips apply in many scenarios. Ensure the thermostat is set properly and the batteries are fresh. Confirm electrical connections are intact, including wiring to the condensing unit. Clean dirt or debris from vents and registers restricting airflow. Clear moisture from window A/C units hindering operation. Replace air filters, as excessive dirt is a top cause of malfunction. Shutting the power off and on again can reset minor control glitches.

When to Seek Professional Help

Seeking professional assistance is advisable if problems involve refrigerant recharging, complex electrical issues, or major component replacement. Most homeowners lack specialized equipment and expertise for involved repairs. Additionally, consult HVAC technicians if basic troubleshooting fails to isolate the issue. They possess the in-depth system knowledge and diagnostic tools to remedy elusive malfunctions.

By cultivating fundamental troubleshooting proficiencies, homeowners can take charge of resolving many HVAC headaches promptly to restore comfort. Correctly identifying issues and solutions prevents wasted energy and costly premature repairs. However, recognize when professional skills are required - don't hesitate to call in an expert. Combining DIY troubleshooting with skilled assistance maximizes cost-savings and keeps your home's HVAC running optimally.

CHAPTER 6
Energy Efficiency in Residential HVAC

Energy-saving Tips

Heating, Ventilation, and Air Conditioning (HVAC) systems are essential components of modern homes, providing comfort throughout the year. However, they can also be significant energy consumers, contributing to high utility bills and environmental impact. In this chapter, we will explore various strategies to enhance energy efficiency in residential HVAC systems.

Regular Maintenance

One of the most fundamental yet often overlooked aspects of maintaining an energy-efficient HVAC system is regular maintenance. Proper maintenance not only extends the lifespan of your system but also ensures it operates at peak efficiency.

Cleaning and Replacing Filters

Filters in your HVAC system play a crucial role in maintaining indoor air quality and system efficiency. Over time, these filters can become clogged with dust, debris, and allergens, causing your system to work harder to maintain the desired temperature. Consequently, it's essential to clean or replace filters regularly, typically every one to three months, depending on your system and local conditions.

Inspecting and Cleaning Ductwork

The ductwork in your home serves as the transport system for conditioned air. Any leaks, blockages, or disconnections in the ducts can result in energy wastage. Regularly inspecting and cleaning your ductwork can ensure that conditioned air reaches its intended destination efficiently.

Professional HVAC Tune-ups

Annual professional HVAC inspections and tune-ups are advisable. These services typically include cleaning coils, lubricating moving parts, checking refrigerant levels, and ensuring the system is calibrated correctly. These preventive measures can catch and resolve issues before they become major problems, saving you money and energy in the long run.

Proper Insulation and Sealing

Effective insulation and sealing are crucial in maintaining a comfortable and energy-efficient home. When your home is inadequately insulated or has air leaks, your HVAC system has to work harder to compensate, leading to higher energy consumption.

Insulation

Ensuring your home is well-insulated is essential. Proper insulation in walls, attics, and floors helps maintain a stable indoor temperature by preventing heat loss in the winter and heat gain in the summer. Common insulation materials include fiberglass, foam board, and cellulose.

Sealing

Air leaks around windows, doors, and other openings in your home can allow conditioned air to escape and outdoor air to infiltrate. Sealing these leaks with weatherstripping, caulk, or spray foam insulation can significantly reduce the load on your HVAC system.

Programmable Thermostats

Installing a programmable thermostat is a smart investment in energy efficiency. These devices allow you to set specific temperature profiles for different times of the day and week, ensuring that your HVAC system only runs when necessary.

Temperature Setbacks

Programmable thermostats enable you to program temperature setbacks when you are not at home or asleep. For example, during the winter, you can set the thermostat to lower the temperature while you're at work and raise it just before you return home. Similarly, during the summer, you can program the thermostat to increase the temperature when you're away.

Smart Thermostats

Smart thermostats take programmability to the next level. They can be controlled remotely via smartphone apps and often have learning capabilities that adapt to your schedule and preferences. Some models can even integrate with other smart home devices for more comprehensive energy management.

Zoning Systems

Zoning systems divide your home into different temperature zones, each with its thermostat. This allows for precise control over the temperature in various areas of your home, reducing energy waste.

How Zoning Works

Zoning systems use dampers in the ductwork to control the flow of conditioned air to different zones. When a particular zone needs heating or cooling, the corresponding damper opens, directing air where it's needed. This prevents over-conditioning of unoccupied spaces.

Benefits of Zoning

Zoning systems offer several benefits, including increased comfort and energy savings. They are particularly useful in larger homes or homes with multiple levels, as they eliminate the need to condition the entire house to meet the comfort needs of a single room.

Energy-efficient Windows and Doors

Windows and doors are critical components of your home's envelope. Inefficient or poorly sealed windows and doors can lead to significant energy losses.

Window Upgrades

Replacing old, single-pane windows with energy-efficient double or triple-pane windows can greatly reduce heat transfer. These windows are designed to prevent heat from escaping during the winter and prevent heat gain during the summer.

Weatherstripping and Sealing

Properly sealing and weatherstripping around windows and doors can eliminate drafts and air leaks, ensuring that your HVAC system isn't working overtime to compensate for these inefficiencies.

Utilize Natural Ventilation

On temperate days, take advantage of natural ventilation to reduce your reliance on mechanical cooling or heating.

Cross-Ventilation

Cross-ventilation involves opening windows on opposite sides of your home to create a flow of fresh air. This can help maintain a comfortable indoor temperature without using your HVAC system.

Night Cooling

In many regions, nighttime temperatures are cooler than daytime temperatures. You can capitalize on this by opening windows at night to allow cool air to circulate through your home. In the morning, close the windows to trap the cool air inside.

Smart Thermostats

In the previous section, we briefly touched on the benefits of programmable and smart thermostats in energy efficiency. In this section, we will delve deeper into how these advanced thermostat systems work and the advantages they offer.

Understanding Smart Thermostats

Smart thermostats are a significant advancement in HVAC technology. They are designed to provide homeowners with more control, convenience, and energy savings compared to traditional thermostats.

Remote Control

One of the primary features of smart thermostats is their ability to be controlled remotely via smartphone apps or web interfaces. This means you can adjust your home's temperature settings from anywhere, ensuring that you only use energy for heating or cooling when it's necessary.

Learning Capabilities

Many smart thermostats come equipped with learning algorithms that analyze your temperature preferences and schedule. Over time, they adapt to your habits, automatically adjusting the temperature to maximize comfort while minimizing energy consumption.

Integration with Other Smart Devices

Smart thermostats often integrate seamlessly with other smart home devices. For example, they can work in tandem with smart lighting systems, motion sensors, and occupancy detectors to further optimize energy usage. For instance, if the thermostat detects that no one is home, it can adjust the temperature accordingly.

Energy Reports and Insights

Smart thermostats provide valuable energy consumption data and insights to help you make informed decisions about your HVAC system usage.

Energy Reports

Many smart thermostat apps generate energy reports that detail your heating and cooling usage. These reports may include information about your energy consumption patterns, the effectiveness of your temperature settings, and suggestions for improvement.

Cost Savings Tracking

Some smart thermostats can also track your energy cost savings. By comparing your energy bills before and after installing the thermostat, you can see the tangible financial benefits of energy-efficient HVAC management.

Compatibility and Installation

Installing a smart thermostat is a relatively straightforward process. However, it's essential to ensure compatibility with your HVAC system before making a purchase.

Compatibility

Smart thermostats are typically compatible with a wide range of HVAC systems, including traditional forced-air systems, heat pumps, radiant heating, and more. Nonetheless, it's vital to check compatibility and, if needed, consult an HVAC professional.

Professional Installation

While some homeowners can install smart thermostats themselves, it's often advisable to have a professional HVAC technician handle the installation. This ensures that the thermostat is correctly wired and calibrated, maximizing its efficiency and effectiveness.

Renewable Energy Options

As society becomes more environmentally conscious, integrating renewable energy sources into our homes has gained considerable momentum. In this section, we will explore renewable energy options for powering HVAC systems and reducing carbon footprints.

Solar Power

Solar power is one of the most popular and accessible renewable energy sources for homeowners. Solar panels installed on the roof or property harness sunlight and convert it into electricity.

How Solar Panels Work

Solar panels use photovoltaic cells to capture sunlight and convert it into direct current (DC) electricity. An inverter then converts the DC electricity into alternating current (AC), which can power your HVAC system and other household appliances.

Solar Battery Storage

To optimize the use of solar power, homeowners can invest in solar battery storage systems. These batteries store excess energy generated during the day for use during nighttime or cloudy periods, allowing for a more consistent and reliable power supply.

Geothermal Heating and Cooling

Geothermal systems utilize the constant temperature of the earth below the surface to provide heating, cooling, and hot water for homes.

How Geothermal Systems Work

Geothermal systems use a heat pump to transfer heat between your home and the ground. During the winter, the system extracts heat from the ground and delivers it to your home. In the summer, it works in reverse, removing heat from your home and transferring it to the ground.

Efficiency and Environmental Benefits

Geothermal systems are highly efficient and environmentally friendly. They use minimal electricity to operate and produce no on-site emissions. Additionally, they can reduce your heating and cooling costs by up to 70%.

Wind Power

For homeowners in areas with suitable wind conditions, wind power can be a viable renewable energy option to power HVAC systems and other household needs.

Residential Wind Turbines

Residential wind turbines, usually mounted on a tower, harness wind energy and convert it into electricity. This electricity can be integrated into your home's electrical system to power your HVAC system and other appliances.

Considerations for Wind Power

Before investing in a residential wind turbine, it's crucial to assess your location's wind patterns and the local regulations regarding wind turbine installations. Wind turbines need consistent and adequate wind speeds to operate efficiently.

PART III
COMMERCIAL HVAC

CHAPTER 7
Scale and Complexity

In the world of HVAC (Heating, Ventilation, and Air Conditioning) systems, understanding scale and complexity is pivotal. This chapter delves deep into the intricacies of these systems, focusing on the differences between residential and commercial HVAC setups, the equipment and components that make them tick, and the critical role of zoning and controls in optimizing performance and energy efficiency.

Differences Between Residential and Commercial

The dichotomy between residential and commercial HVAC systems is profound, extending beyond mere differences in scale. While both share the fundamental objective of maintaining indoor comfort, they diverge significantly in complexity, design, regulation, and purpose. Unraveling these distinctions is vital for a comprehensive comprehension of HVAC systems and their tailored applications.

Scale and Purpose

The most evident disparity between residential and commercial HVAC systems is their scale and purpose. Residential HVAC systems are engineered to cater to the heating, ventilation, and cooling needs of individual homes or small apartments. Typically, these systems serve areas encompassing less than 5,000 square feet. In essence, the focus is on personalized comfort and efficiency, ensuring the well-being of a relatively limited number of occupants.

On the contrary, commercial HVAC systems undertake the monumental task of conditioning larger spaces, ranging from small retail stores to extensive warehouses, corporate offices, and towering skyscrapers. The scale of these systems can encompass thousands to millions of square feet, accommodating a multitude of occupants and varying usage patterns. Consequently, commercial HVAC systems are engineered with a broader, strategic perspective, aiming to maintain a conducive environment for both individuals and business operations.

Design Complexity and Installation

Design complexity is another fundamental contrast. Residential HVAC systems are designed with simplicity and ease of installation in mind. They often comprise a single furnace or air conditioner, a basic thermostat, and a network of ducts or vents distributing air throughout the living space. Due to their relatively uncomplicated design, installation is typically quicker and more straightforward.

On the flip side, commercial HVAC systems entail intricate planning, engineering, and installation processes. The design integrates a multitude of units, advanced control systems, complex ductwork, and specialized equipment such as chillers, boilers, and rooftop units. This complexity is essential to efficiently manage a vast and diverse space, ensuring uniform temperature distribution and ventilation across the entire commercial establishment.

Regulatory Environment

The regulatory environment is a significant differentiator between residential and commercial HVAC systems. Residential systems are subject to local building codes and regulations, but the scope and stringency of these requirements are usually less compared to their commercial counterparts.

In the commercial realm, compliance with an intricate web of codes and standards is imperative. These encompass safety regulations, environmental norms, and stringent energy efficiency guidelines. The intricacy of these regulations necessitates a higher level of expertise during the design, installation, and ongoing maintenance of commercial HVAC systems. Failure to comply can result in severe penalties and disruptions in business operations.

Energy Efficiency and Sustainability

Energy efficiency and sustainability concerns also diverge significantly between residential and commercial systems. Residential systems emphasize energy efficiency to reduce energy bills for homeowners and minimize their environmental footprint. Energy-efficient appliances, better insulation, and programmable thermostats are common strategies in the residential sphere.

In the commercial sector, energy efficiency is a paramount consideration due to the massive energy consumption associated with larger spaces. Additionally, businesses often face stringent energy usage targets and sustainability goals set by regulatory bodies or as part of corporate responsibility initiatives. As a result, commercial HVAC systems often employ advanced technologies and control mechanisms to optimize energy consumption while maintaining optimal indoor conditions.

Maintenance and Service

The maintenance and servicing of residential and commercial HVAC systems also exhibit noteworthy differences. In residential settings, maintenance is usually managed by homeowners or contracted to local HVAC service providers. The maintenance routines are more standardized and typically involve tasks like changing filters, cleaning ducts, and conducting periodic system check-ups.

Conversely, the maintenance of commercial HVAC systems is more intricate and demanding. Due to the scale and complexity of these systems, specialized technicians and maintenance teams are employed to handle the periodic upkeep. Regular inspections, complex repairs, and the calibration of advanced control systems are critical aspects of maintaining commercial HVAC systems at peak performance. Additionally, commercial systems often have maintenance contracts in place, ensuring timely and expert attention to the system's needs.

Equipment and Components

The world of Heating, Ventilation, and Air Conditioning (HVAC) is an intricate network of equipment and components that work in harmony to provide comfort and climate control to both residential and commercial spaces. Understanding these components is essential for anyone involved in the HVAC industry, from technicians and engineers to homeowners seeking a deeper comprehension of their HVAC systems.

Heating and Cooling Units

The heart of an HVAC system lies in its heating and cooling equipment. In residential setups, a furnace is commonly used for heating purposes, generating warm air that is then circulated throughout the living space. On the cooling side, air conditioners or heat pumps take center stage, pulling warm air from inside and releasing cooler air.

In commercial settings, the equipment tends to be more diverse to accommodate larger spaces and varied requirements. Boilers, for instance, are often utilized for heating in commercial HVAC systems. These appliances use water or steam to distribute heat through radiators or other devices. Chillers, on the other hand, are a crucial part of the cooling process. They work by removing heat from the indoor air and transferring it to the outside environment. Rooftop units, which combine heating and cooling functions, are also prevalent in commercial installations, especially in mid-sized to large buildings.

Ductwork and Vents

Ductwork and vents form the respiratory system of an HVAC setup, ensuring the proper distribution of conditioned air throughout the building. In residential systems, ducts are typically less complex, consisting of a single set that branches off to various rooms or zones. The vents, strategically placed, deliver heated or cooled air to these spaces.

In contrast, commercial HVAC systems necessitate a more intricate ductwork layout due to the larger and more complex nature of the buildings they serve. The ducts are usually more substantial, well-insulated, and designed to accommodate the higher airflow demands of larger spaces. Additionally, dampers and valves may be integrated into the ductwork to control and balance the airflow, ensuring consistent heating and cooling levels across all areas.

Thermostats and Controls

Thermostats and controls are like the brains of an HVAC system, translating the occupants' comfort preferences into specific actions of heating, cooling, or maintaining a set temperature. In residential setups, thermostats are often simple devices that allow users to set the desired temperature manually. With the advent of smart technology, however, many residential thermostats now offer programmable features and can be controlled remotely through mobile applications.

On the commercial front, control systems are more sophisticated, given the complexity and scale of the HVAC setups. Building Automation Systems (BAS) take the lead in commercial control. These systems manage and control the overall climate of a building, allowing for precise regulation of temperature, humidity, and ventilation. Advanced sensors and controls in a BAS continuously collect data, enabling real-time adjustments for optimal energy efficiency and comfort.

Filters and Air Quality Components

Indoor air quality (IAQ) is an increasingly significant concern in both residential and commercial spaces. Filters and air quality components are fundamental in addressing this issue. Residential systems typically use standard filters that capture dust, pollen, and other particles to ensure the air circulated within the home is clean.

In contrast, commercial HVAC systems often require more sophisticated filtration systems to meet stringent IAQ standards. High-efficiency particulate air (HEPA) filters and activated carbon filters are commonly employed to trap even finer particles and remove odors and harmful gases from the air. Additionally, ultraviolet (UV) germicidal lamps may be integrated into commercial systems to disinfect the air, particularly in healthcare facilities and sensitive environments.

Zoning and Controls

Zoning and controls represent a pivotal aspect of HVAC system design, especially in larger commercial buildings where precise control and energy efficiency are paramount. These elements offer the ability to customize and manage the indoor environment according to the diverse needs of different zones within a building.

Zoning

Zoning involves dividing a building into distinct areas, known as zones, based on factors such as occupancy, usage, and temperature preferences. This process allows for individualized control of the heating, cooling, and ventilation for each zone. In residential applications, zoning is relatively straightforward and is often achieved by using multiple thermostats that control dampers or ductless mini-split systems for different areas of the house.

In a commercial setup, the zoning concept is significantly more complex due to the size and layout of the building. Different areas, such as offices, meeting rooms, or open workspaces, may have diverse heating and cooling needs. Therefore, advanced control systems are used to regulate the HVAC equipment serving each zone, optimizing comfort and energy usage. This can result in substantial energy savings by ensuring that only the areas in use receive conditioned air.

Control Systems

Control systems are the central nervous system of an HVAC setup, directing the operation of various components to achieve the desired indoor conditions. Residential control systems are relatively basic, often involving a single thermostat that communicates with the heating and cooling equipment. Modern advancements have led to the development of smart thermostats, which can be programmed and controlled remotely through smartphones or other devices.

On the commercial side, control systems are more intricate, and Building Automation Systems (BAS) is a common choice. BAS integrates numerous components, including sensors, controllers, and software, to automate HVAC operations and provide actionable data for building management. The controls can be centralized or decentralized, allowing for precise control and monitoring of HVAC parameters for each zone or even individual rooms. These systems also enable scheduling, setpoint optimization, and adaptive control strategies, enhancing energy efficiency and overall system performance.

Energy Management

Energy management is a critical concern in both residential and commercial HVAC systems. However, due to the larger scale and higher energy consumption in commercial buildings, energy management takes on a more significant role in the commercial context. Commercial systems often incorporate energy-efficient components and technologies to minimize energy usage while maintaining comfort levels.

Variable Frequency Drives (VFDs) are a prime example. They are frequently used in commercial HVAC systems to control the speed of fans and pumps. By adjusting the speed based on the actual demand, VFDs save energy compared to traditional on-off controls, which often operate at full speed even when not needed.

Demand-controlled ventilation is another energy-saving strategy utilized in commercial HVAC. This technology adjusts the outside air intake based on the occupancy levels within the building. During low occupancy periods, the system reduces the amount of outside air brought in, saving on heating or cooling energy.

Maintenance and Monitoring

Maintenance and monitoring are indispensable aspects of HVAC systems to ensure their longevity, efficiency, and safety. In residential settings, homeowners often undertake basic maintenance tasks themselves, such as replacing air filters regularly. Occasionally, they may schedule professional HVAC technicians for comprehensive check-ups.

In the commercial realm, due to the scale and complexity of the systems, maintenance is a more elaborate and specialized affair. HVAC technicians conduct routine inspections, cleaning, and maintenance of critical components. Additionally, many commercial HVAC systems are equipped with remote monitoring capabilities. These features allow facility managers to keep a close eye on system performance, detect anomalies, and schedule timely maintenance to prevent potential breakdowns.

Building Automation Systems (BAS)

Building Automation Systems (BAS) are the epitome of sophisticated HVAC control in commercial settings. These systems integrate various components and technologies to automate and optimize a building's climate control, lighting, security, and other systems.

In the HVAC realm, a BAS orchestrates the operation of heating, cooling, and ventilation based on data from sensors placed throughout the building. These sensors provide real-time information about temperature, occupancy, humidity, and more. The BAS processes this data and adjusts HVAC settings accordingly, aiming for maximum energy efficiency while maintaining the desired comfort levels.

BAS often have user-friendly interfaces, enabling facility managers to monitor and control the HVAC system from a central control panel. They can set schedules, adjust setpoints, and receive alerts for maintenance or system malfunctions, streamlining the management and operation of the entire HVAC system.

Challenges in Commercial Zoning and Controls

While commercial zoning and control systems offer numerous benefits, they come with their set of challenges. One of the primary challenges lies in the complexity of these systems. Commercial buildings are vast and house numerous zones, each with its unique heating and cooling requirements. Designing a zoning system that caters to these diverse needs while ensuring efficient operation can be a daunting task.

Moreover, user interfaces in commercial control systems can be intimidating for non-experts. Facility managers and building occupants need to be adequately trained to utilize these systems effectively. A clear and intuitive interface design is crucial to ensure that users can navigate the controls with ease.

Security is also a growing concern, especially with the integration of digital technologies in HVAC control. Protecting the BAS and the overall system from cyber threats is paramount to prevent unauthorized access, data breaches, or potential disruptions to the building's operations.

A deep understanding of HVAC equipment, components, zoning, and control systems is indispensable for anyone involved in the HVAC industry. From the simplicity of residential setups to the complexity of commercial systems, each component and control mechanism plays a vital role in creating a comfortable, energy-efficient indoor environment. Keeping pace with advancements, addressing challenges, and embracing sustainable practices will shape the future of HVAC systems, ensuring optimal performance and enhanced quality of life for all.

CHAPTER 8
Specialized Applications

In our exploration of HVAC systems, we've already covered the basics, including how they work, their components, and considerations for residential and commercial spaces. As we delve deeper into this subject, we reach a critical juncture where we explore specialized applications of HVAC systems. These applications are tailored to meet the unique needs of specific industries and environments. In this chapter, we'll examine three specialized applications: HVAC for Healthcare, HVAC for Manufacturing, and HVAC in Office Spaces. Each of these sectors demands a nuanced approach to heating, ventilation, and air conditioning to ensure optimal comfort, safety, and efficiency.

HVAC for Healthcare

Healthcare facilities, such as hospitals and clinics, present a unique set of challenges when it comes to HVAC systems. The environment within healthcare facilities plays a pivotal role in patient comfort, safety, and recovery, as well as the well-being of healthcare professionals. The following section delves into the intricacies of HVAC systems in healthcare settings.

The Importance of HVAC in Healthcare

Healthcare facilities, by their nature, have distinct requirements for indoor air quality (IAQ) and environmental control. These requirements are critical for several reasons:

Infection Control

Hospitals are at the forefront of battling infectious diseases. HVAC systems must minimize the risk of airborne transmission by maintaining the right temperature and humidity levels and controlling airflows to ensure that contaminated air doesn't spread.

Comfort for Patients and Staff

Patient comfort is vital for the healing process, and healthcare staff need a comfortable working environment to provide the best care. HVAC systems must provide individualized control in patient rooms and maintain optimal conditions in common areas and staff spaces.

Ventilation

Proper ventilation is essential for removing pollutants and replenishing oxygen. In healthcare settings, this is crucial for reducing the risk of airborne infections and ensuring that patients and staff are breathing clean air.

Regulatory Compliance

Healthcare facilities are subject to strict regulations, including those set forth by organizations like the Joint Commission. HVAC systems must meet these regulations to maintain accreditation.

HVAC System Components for Healthcare

To meet the unique needs of healthcare facilities, HVAC systems incorporate several specialized components and features:

Air Filtration

Hospitals require high-efficiency air filters to capture particles and pathogens. HEPA (High-Efficiency Particulate Air) filters are commonly used to ensure the air is clean and safe.

Zoning

Zoning allows for different temperature settings in various areas of the facility. Patient rooms, operating rooms, and waiting areas may all have different requirements, and zoning ensures each area gets the right conditioning.

Air Changes per Hour (ACH)

ACH measures how many times the air in a room is replaced with fresh air per hour. In healthcare settings, a high ACH is often necessary to ensure that contaminants are continually removed from the air.

Humidity Control

Maintaining the right humidity levels is critical for infection control and patient comfort. Humidity levels that are too high can encourage mold growth, while levels that are too low can cause discomfort and increase the risk of infection.

Energy Efficiency and Sustainability in Healthcare HVAC

In recent years, there has been a growing focus on energy efficiency and sustainability in healthcare HVAC systems. Healthcare facilities are among the most energy-intensive buildings due to their 24/7 operation. Implementing energy-efficient HVAC solutions not only reduces operating costs but also aligns with the broader environmental goals of reducing carbon emissions. Some strategies for achieving energy efficiency in healthcare HVAC include:

1. **Variable Speed Drives: These controls adjust fan and pump speeds to match the current load, reducing energy consumption during low-demand periods.**
2. **Heat Recovery: Heat recovery systems capture waste heat from various HVAC processes and**

use it to preheat incoming air or water, improving overall efficiency.

3. **Building Automation:** Smart HVAC systems that use sensors and predictive algorithms can optimize HVAC performance in real-time, adjusting settings based on occupancy and outdoor conditions.

4. **Regular Maintenance:** Routine maintenance and cleaning of HVAC equipment ensure it operates at peak efficiency.

HVAC for Manufacturing

Within manufacturing facilities, HVAC systems perform a vital role in maintaining consistent indoor conditions to support high-quality processes and worker productivity. The scale and complexity of manufacturing HVAC make optimizing these systems crucial for efficiency.

Manufacturing HVAC differs fundamentally from the comfort conditioning of offices or homes due to the need for precision process control and contaminant management. Tight regulation of parameters like temperature and humidity ensures consistent product output, while specialized ventilation removes problematic dust and fumes inherent in industrial processes. Worker safety and comfort remain a priority as well, requiring attention to be paid to air quality and moderating work area temperatures. Additionally, manufacturing machinery generates substantial heat that must be dissipated to prevent disruptive equipment failures.

Clean rooms epitomize the stringent demands placed on manufacturing HVAC. Semiconductor fabrication, pharmaceutical production, and other processes producing highly sensitive products require nearly particle-free air with minimal deviations in conditions. To achieve this ultra-clean state, clean rooms integrate advanced HEPA filtration, laminar airflow diffusers, humidity control, specialized pressure balancing, and redundancy in critical equipment.

Ventilation rates in manufacturing areas often far exceed commercial settings to dilute and remove air contaminants. Large air handling units with high-capacity exhaust fans exchange air rapidly. Supply systems temper and filter this fresh air before distribution. Sophisticated dust collection systems target capturing emissions directly at the source to augment general ventilation.

Supporting specialized industrial processes adds further complexities. Food manufacturing refrigeration precisely controls temperatures for proper freezing, thawing, and preparation. Plastic production chiller systems remove heat from molding equipment using chilled water or other liquid coolant loops. Textile facility humidity control maintains proper moisture levels for scouring, dying, and finishing operations.

With heavy usage, optimizing manufacturing HVAC efficiency is imperative yet challenging. Utilizing variable-speed drives, high-efficiency chillers, and heat recovery wheels makes incremental improvements. Centralizing heating and cooling in larger facilities allows for capturing waste heat for reuse. Advanced control systems help match ventilation airflow to real-time contaminant levels rather than over-ventilating perpetually. Regular maintenance ensures peak performance across expansive HVAC installations.

HVAC in Office Spaces

Within office buildings, HVAC systems carry the important task of maintaining comfortable and healthy indoor environments for workers over long hours of occupation. Balancing temperature control, air quality, operating costs, and zoning flexibility is imperative for productivity.

Office HVAC differs from residential applications in several key aspects. Temperature ranges must support sedentary, long-duration occupancy across diverse climates. Ventilation rates need to sufficiently dilute occupant emissions and filter particulates from large volumes of recirculated air. Extended daily operation translates to maximizing energy efficiency to control expenses. And most importantly, zoning and controls must allow customizing conditions across the many independently occupied rooms of multi-tenant offices.

VAV systems shine in office applications by varying air volumes delivered based on real-time occupancy and loads. Reducing airflow to vacant offices saves substantial fan energy while sustaining comfort in populated zones. Occupancy sensors further optimize efficiency by signaling VAV boxes to reduce ventilation when workers leave. Thermal storage shifts cooling to off-peak hours when electricity rates are lower, while daylight harvesting adjusts artificial lights and related heating/cooling based on available sunlight.

Complimenting the HVAC system itself, smart thermostats reduce heating and cooling waste dramatically through adjustable setback periods that match office schedules. Building automation systems monitor and coordinate equipment operation enterprise-wide to ensure cohesive optimization. Proper maintenance keeps components operating at peak efficiency to minimize energy waste. Promoting energy-conscious behaviors among office workers, such as turning off unnecessary lights and closing blinds during peak sun, further augments savings.

CHAPTER 9
Commercial HVAC Maintenance and Regulations

In the realm of commercial buildings, maintaining an efficient and smoothly running HVAC (Heating, Ventilation, and Air Conditioning) system is paramount. HVAC systems are a vital component of any commercial establishment, ensuring a comfortable environment for employees, customers, and the overall functionality of the business. This chapter delves into the critical aspects of commercial HVAC maintenance and the legal and regulatory guidelines that govern its operations.

Preventative Maintenance

In commercial HVAC (Heating, Ventilation, and Air Conditioning) systems, maintaining an efficient and smoothly running infrastructure is a critical aspect of operational success. A proactive approach to this maintenance is known as preventative maintenance. In essence, preventative maintenance encompasses a series of scheduled inspections, tasks, and repairs aimed at preventing system breakdowns, enhancing system efficiency, extending equipment life, and minimizing operational costs. This section delves into the importance, components, and benefits of preventative maintenance in commercial HVAC systems.

Importance of Preventative Maintenance

Cost Savings and Operational Efficiency

Preventative maintenance significantly contributes to cost savings and operational efficiency. Regularly scheduled check-ups and maintenance tasks can identify potential issues early on, preventing major system failures that could result in expensive repairs or even complete system replacements. In the long run, the cost of performing routine maintenance is significantly less than dealing with unexpected breakdowns.

Extended Equipment Life

Just as regular exercise and a balanced diet contribute to a longer, healthier life for humans, routine maintenance extends the lifespan of HVAC equipment. Regular cleaning, lubrication, and minor repairs ensure that the system operates smoothly and as intended, delaying the need for major overhauls or replacements.

Enhanced Energy Efficiency

An efficiently running HVAC system consumes less energy, leading to reduced utility bills. Preventative maintenance optimizes system performance by ensuring that all components work at their best capacity, minimizing energy wastage, and keeping costs in check.

Improved Indoor Air Quality

Regular filter replacements and cleaning as part of preventative maintenance contribute to better indoor air quality. Filters that are clogged with dust and debris can lead to poor air circulation and compromised air quality, which could have adverse effects on the health and well-being of building occupants.

Components of Preventative Maintenance

Regular Inspections

Regular inspections are the cornerstone of preventative maintenance. Trained technicians conduct detailed evaluations of the HVAC system, checking for any signs of wear and tear, leakages, airflow issues, and potential problem areas. These inspections are usually scheduled at specific intervals, allowing for timely identification and resolution of emerging issues.

Cleaning and Lubrication

Dust, dirt, and debris are natural adversaries of HVAC systems. These particles can accumulate in various components, hindering optimal performance. Routine cleaning ensures that these elements are removed, and lubrication of moving parts reduces friction, enhancing overall efficiency and extending the life of the equipment.

Filter Replacement

Air filters play a pivotal role in maintaining air quality within a building. Over time, these filters accumulate dirt and pollutants, leading to reduced airflow and diminished air quality. Routine replacement of filters is crucial to prevent clogging and maintain optimal airflow and air quality.

Calibration of Controls

The proper functioning of HVAC systems heavily relies on accurate readings and control settings. Through regular maintenance, the controls and thermostats are calibrated, ensuring that they provide precise readings and enabling the system to operate at its most efficient levels.

Benefits of Preventative Maintenance

Operational Reliability

Preventative maintenance instills confidence in the reliability of the HVAC system. When a system is well-maintained, the likelihood of unexpected breakdowns during critical periods is significantly reduced, ensuring consistent and uninterrupted comfort for building occupants.

Enhanced Safety

Safety is a critical consideration in any commercial establishment. A well-maintained HVAC system minimizes potential safety hazards associated with malfunctioning equipment, such as electrical issues or gas leaks, ensuring the safety of employees and visitors.

Sustainable Practices

In a world increasingly focused on sustainability, preventative maintenance aligns with environmentally responsible practices. By optimizing the efficiency of HVAC systems and extending their lifespan, businesses reduce their environmental footprint by decreasing the resources required for manufacturing and disposing of new equipment.

Maintaining Warranty Compliance

Many HVAC systems come with warranties that require regular maintenance to remain valid. Adhering to a preventative maintenance schedule ensures compliance with warranty terms, potentially saving on repair or replacement costs covered by the warranty.

Preventative maintenance is the linchpin of a successful commercial HVAC system. It not only contributes to significant cost savings and operational efficiency but also extends the life of the equipment and fosters a safer and healthier indoor environment. Businesses that invest in comprehensive preventative maintenance programs for their HVAC systems reap the benefits of enhanced energy efficiency, improved air quality, and a sustainable operational approach—all of which are vital for long-term success and responsible business practices.

Energy Management Systems

In today's world, where energy conservation and efficiency are of utmost importance, Energy Management Systems (EMS) have emerged as a vital tool for businesses, especially in the realm of HVAC operations within commercial buildings. These systems play a pivotal role in not only optimizing energy consumption but also in reducing operational costs and promoting sustainable practices.

Understanding Energy Management Systems (EMS)

Energy Management Systems are comprehensive frameworks that oversee, monitor, and optimize energy-consuming assets and processes within a commercial establishment. The primary goal is to ensure that energy is utilized judiciously and efficiently while maintaining the desired levels of performance and comfort.

Components of Energy Management Systems

Sensors and Meters

A cornerstone of any EMS is its ability to gather precise and timely data. This is achieved through an array of sensors and meters strategically placed within the building. These sensors can monitor a variety of parameters, including temperature, humidity, occupancy, and energy usage. Sophisticated meters record energy consumption across different sections of the establishment, providing valuable insights into usage patterns.

Control Algorithms

EMS employs advanced algorithms that process the data collected by the sensors and meters. These algorithms are designed to analyze the data and make real-time adjustments to the HVAC systems and other energy-consuming components. The adjustments are based on a set of predefined parameters and may include actions like regulating temperature, adjusting lighting levels, or controlling the HVAC system's functioning.

Human-Machine Interface (HMI)

The Human-Machine Interface (HMI) is the bridge between the system and its operators. It offers a user-friendly platform that allows individuals to interact with the EMS. The HMI provides real-time information on energy consumption, system performance, and other relevant data. It also permits operators to set parameters, make adjustments, and generate reports for further analysis.

Benefits of Energy Management Systems

Energy Savings

One of the most significant advantages of implementing an EMS is the potential for substantial energy savings. By monitoring energy consumption patterns and making real-time adjustments, the system ensures that energy is used only when necessary. This optimized usage directly translates to lower energy bills for the business.

Cost Savings

Lower energy consumption naturally results in reduced operational costs. The cost savings can be significant, especially for large commercial establishments where energy bills constitute a substantial portion of the operational expenses. These saved funds can be redirected to other essential areas of the business.

Environmental Sustainability

Energy Management Systems contribute to environmental sustainability by reducing overall energy consumption. With a lower carbon footprint, businesses can align with environmental regulations and demonstrate their commitment to responsible corporate citizenship.

Predictive Maintenance

EMS often includes predictive maintenance features. By continuously monitoring equipment and analyzing data, the system can predict when maintenance is required. This proactive approach to maintenance helps prevent major breakdowns and prolongs the lifespan of HVAC systems and other equipment.

Enhanced Productivity

A well-controlled, comfortable environment resulting from efficient HVAC operations can lead to increased productivity among the occupants of the commercial space. When individuals are comfortable, they are more focused and productive in their tasks.

Challenges and Considerations

Implementing an EMS is not without its challenges. The initial setup cost can be significant, often acting as a deterrent for small or medium-sized businesses. Moreover, integrating an EMS into an existing infrastructure may require modifications and upgrades, which can be time-consuming and disruptive to regular operations. However, the long-term benefits far outweigh the initial challenges.

Another critical consideration is cybersecurity. With the integration of digital interfaces and networks, EMS becomes vulnerable to cyber threats. Hence, ensuring a robust cybersecurity framework is essential to safeguard sensitive data and prevent unauthorized access.

Legal and Regulatory Guidelines

In the dynamic landscape of commercial HVAC systems, compliance with legal and regulatory guidelines is imperative. These guidelines are established and enforced by various regulatory bodies to ensure the safe, efficient, and environmentally responsible operation of HVAC systems within commercial buildings.

Regulatory Bodies

Several regulatory bodies are instrumental in defining and enforcing the guidelines that govern commercial HVAC systems.

Occupational Safety and Health Administration (OSHA)

OSHA, a federal agency in the United States, is primarily concerned with ensuring the safety and health of workers. In the context of HVAC systems, OSHA sets and enforces safety standards and regulations to protect workers who operate, maintain, and work around HVAC equipment.

For instance, OSHA regulations dictate proper ventilation and air quality standards within workplaces to ensure the well-being of employees.

Environmental Protection Agency (EPA)

The EPA, another federal agency in the U.S., focuses on environmental regulations. Concerning HVAC systems, the EPA governs the use and handling of refrigerants, aiming to reduce the use of harmful substances that can have adverse effects on the environment.

The EPA's regulations include guidelines for proper refrigerant management, leak detection, and responsible disposal practices.

American Society of Heating, Refrigerating, and Air Conditioning Engineers (ASHRAE)

ASHRAE, a professional association, develops industry standards and guidelines for HVAC systems. These standards cover a wide array of aspects, including system design, energy efficiency, indoor air quality, and sustainability.

Businesses often adhere to ASHRAE standards to ensure their HVAC systems meet the latest industry benchmarks for efficiency and safety.

Compliance and Penalties

Compliance with these regulatory guidelines is mandatory for commercial establishments. Non-compliance can have severe consequences, ranging from fines and legal liabilities to potential closure of the business.

Fines and Penalties

Violations of regulatory guidelines can lead to fines of varying amounts, depending on the severity of the violation. The fines can accumulate and become a substantial financial burden for the business, impacting its operational budget.

Legal Liabilities

Non-compliance with safety and environmental regulations can expose the business to legal liabilities. In case of accidents or health issues related to HVAC systems, the business may face lawsuits, resulting in further financial and reputational damage.

Forced Closure

In extreme cases where repeated or severe violations persist, regulatory bodies have the authority to order the business to halt operations until compliance is achieved. This can lead to significant financial losses, business disruptions, and a damaged reputation in the market.

Importance of Compliance

Adhering to legal and regulatory guidelines is not only a legal requirement but also a mark of responsible and ethical business conduct. It ensures the safety and well-being of occupants, contributes to environmental sustainability, and upholds the reputation and credibility of the business.

Moreover, compliance showcases the commitment of a business to uphold high standards, thereby fostering trust among clients, partners, and the community. It also positions the business as a responsible entity that takes its social and environmental responsibilities seriously.

PART IV
ADVANCED TOPICS AND TECHNOLOGIES

CHAPTER 10
Emerging Trends

In the rapidly evolving landscape of technology and innovation, the realm of Heating, Ventilation, and Air Conditioning (HVAC) is undergoing significant transformations. The convergence of digitalization and climate control has given birth to revolutionary concepts, propelling the industry into a new era. This chapter delves deep into three groundbreaking trends that are shaping the future of HVAC systems: Smart HVAC Systems, IoT in HVAC, and Future Predictions.

Smart HVAC Systems

The contemporary world is in the midst of a technological renaissance, and this wave of innovation has significantly impacted the field of Heating, Ventilation, and Air Conditioning (HVAC). Within this realm, the concept of Smart HVAC systems has emerged as a beacon of efficiency and sustainability. These cutting-edge systems amalgamate state-of-the-art sensors, artificial intelligence (AI), and automation to optimize heating, cooling, and ventilation processes in various environments. The ultimate objective is to provide a comfortable indoor climate while minimizing energy usage and subsequently reducing the overall carbon footprint.

Components and Architecture

The architecture of Smart HVAC systems is a symphony of intricate components working harmoniously to achieve optimal performance and energy efficiency.

Sensors and Actuators

At the core of these systems are an array of sensors strategically placed throughout the building. These sensors gather a vast array of data, including temperature, humidity levels, occupancy patterns, and air quality. Actuators, on the other hand, respond to the data collected by the sensors, adjusting heating, cooling, and airflow systems accordingly. For instance, if a room is unoccupied, the system can automatically adjust the temperature to save energy.

Control Units

The heart and brain of the Smart HVAC system reside within sophisticated control units. These units are often powered by advanced AI algorithms capable of interpreting the data collected by sensors in real-time. Based on this analysis, the system makes intelligent decisions to optimize HVAC operations. These decisions range from minor adjustments in thermostat settings to regulating the speed of fans and compressors, all with the aim of achieving the ideal indoor climate efficiently.

Connectivity

Connectivity is the sinew that allows these systems to operate seamlessly. Smart HVAC systems are intertwined with the internet, enabling them to be monitored and controlled remotely. This is a revolutionary aspect of these systems as it grants users the power to manage their HVAC settings through dedicated applications or web interfaces, providing unparalleled convenience and control over the indoor climate.

Benefits

The integration of intelligence into HVAC systems offers a plethora of benefits that revolutionize how we manage our indoor environments.

Energy Efficiency

Smart HVAC systems excel in energy efficiency. By analyzing usage patterns and adjusting settings in real-time, they can significantly reduce energy consumption. For example, during non-peak hours or when a space is unoccupied, the system can automatically adjust temperature settings to minimize energy usage, thus leading to substantial cost savings.

Cost Savings

Reduced energy consumption directly translates into cost savings, making these systems economically viable in the long run. The initial investment, though higher, pays for itself over time through lower energy bills.

Enhanced Comfort and Health

Smart HVAC systems offer unparalleled comfort to the occupants of a space. By tailoring HVAC settings according to individual preferences, they ensure a more comfortable indoor climate. Moreover, these systems often lead to improved air quality by efficiently ventilating spaces, contributing to a healthier living or working environment.

Remote Monitoring and Control

One of the standout features of Smart HVAC systems is the ability for remote monitoring and control. Whether from a different part of the building or from a different city, users can easily manage and adjust their HVAC systems through dedicated applications on their smartphones or via web portals. This is particularly beneficial for large commercial buildings or businesses with multiple locations.

Challenges

While the advantages of Smart HVAC systems are evident, they are not without their set of challenges.

Initial Costs

The upfront cost of implementing a Smart HVAC system can be a significant deterrent for potential adopters. The initial investment required for sensors, control units, and connectivity features can be substantial, potentially limiting its adoption, especially for smaller businesses or residential applications.

Complexity

The complexity of Smart HVAC systems necessitates specialized knowledge for installation, maintenance, and troubleshooting. Not every technician is adept at handling the intricacies of these systems, which can pose a challenge for both users and technicians.

Privacy and Security Concerns

With the influx of interconnected systems and data transmission, privacy and security concerns arise. The collection of data on usage patterns and personal preferences raises questions about how this data is stored, who has access to it, and how it can be potentially misused. Robust data security measures must be in place to address these concerns.

Smart HVAC systems epitomize the convergence of cutting-edge technology and sustainability in the realm of climate control. As advancements continue to refine these systems and drive down initial costs, they are poised to become the standard in the HVAC industry. The tremendous potential for energy savings, cost efficiency, and enhanced comfort makes these systems a pivotal innovation for a greener and more comfortable future.

IoT in HVAC: Revolutionizing Climate Control

The convergence of the Internet of Things (IoT) and the realm of Heating, Ventilation, and Air Conditioning (HVAC) systems is ushering in a new era of climate control. IoT integration in HVAC involves connecting devices, sensors, and systems to a centralized network, enabling seamless communication and data sharing. This interconnectedness brings forth transformative possibilities, from predictive maintenance to energy efficiency, significantly impacting the way HVAC systems are operated and maintained.

Integration and Functionality

The integration of IoT in HVAC is a sophisticated process that involves several components and functionalities.

a) Sensors

Sensors are the bedrock of IoT integration in HVAC systems. These devices are strategically placed within HVAC systems to capture a myriad of data. Temperature sensors monitor the ambient temperature, humidity sensors measure the moisture levels in the air, and air quality sensors assess pollutants like carbon dioxide, particulate matter, and volatile organic compounds. Furthermore, sensors are embedded in HVAC equipment to monitor their performance and operational parameters.

b) Connectivity

IoT-enabled HVAC systems rely on various communication protocols and mediums to transmit data. Wi-Fi, Bluetooth, Zigbee, and LoRaWAN are common connectivity options that facilitate communication between sensors, actuators, and centralized platforms. This connectivity allows for real-time data transmission, enabling immediate response and action based on the gathered data.

c) Data Analysis

Once data is collected from the myriad of sensors, advanced analytics, and machine learning algorithms process and analyze this data. These analyses help in identifying patterns, trends, and anomalies, which in turn can be used to optimize HVAC system performance, predict maintenance needs, and enhance energy efficiency.

d) Centralized Control Platforms

The data and insights gathered from sensors are typically sent to centralized platforms, often cloud-based, for further analysis and control. These platforms provide a comprehensive view of HVAC system performance and can be accessed remotely, allowing for real-time monitoring and control. They facilitate decision-making by providing actionable insights to optimize the HVAC system's operation.

Advantages

The integration of IoT in HVAC systems brings about a host of advantages that have a profound impact on efficiency, cost-effectiveness, and overall user experience.

a) Predictive Maintenance

One of the significant advantages of IoT in HVAC is predictive maintenance. Through continuous monitoring and analysis of data from sensors embedded within the HVAC system, the system can predict when maintenance is required. This proactive approach prevents unexpected breakdowns, reduces downtime, and saves on maintenance costs.

b) Energy Efficiency

Real-time data analysis allows for immediate adjustments in the HVAC system based on current needs. For example, the system can adjust temperature settings based on occupancy, ensuring energy is used efficiently. These energy-saving measures can significantly reduce operational costs and contribute to a greener, more sustainable environment.

c) Improved Performance

Constant monitoring and data-driven adjustments ensure that HVAC systems operate at peak performance levels consistently. This leads to a more comfortable indoor environment for users while optimizing the system's performance and longevity.

d) Enhanced User Experience

The ability for users to control and monitor their HVAC systems remotely through mobile apps or web interfaces is a game-changer. It provides a seamless and personalized experience, allowing users to tailor their climate preferences and schedules according to their needs.

Challenges

However, the integration of IoT in HVAC systems is not without its challenges, which need to be addressed for seamless implementation and adoption.

a) Data Privacy and Security

The gathering and transmission of data raise concerns about privacy and security. HVAC systems hold sensitive data about the building and its occupants. Therefore, robust cybersecurity measures must be in place to protect this data from potential breaches.

b) Interoperability

Interoperability refers to the seamless communication and compatibility between various devices and systems from different manufacturers. Achieving interoperability is crucial to ensure that all components within the IoT-enabled HVAC system can work together effectively.

c) Scalability

IoT implementations need to be scalable to accommodate the increasing number of connected devices and the associated data volume as the system expands. Ensuring scalability is essential to maintain optimal system performance.

Future Predictions

The world of HVAC is evolving rapidly, driven by technological innovation, sustainability imperatives, and shifting societal needs. Predicting the future of HVAC involves envisioning a trajectory where technology intertwines with sustainability, efficiency, and user-centricity. In this section, we explore the potential future trends and advancements that are set to shape the HVAC industry in the coming years.

Green HVAC Technologies

A prominent trajectory in the future of HVAC systems is undeniably green. Energy efficiency and environmental sustainability are now paramount in HVAC advancements. The imperative to reduce carbon emissions and mitigate the impacts of climate change is pushing the HVAC industry to embrace greener technologies. Green HVAC technologies encompass a range of approaches, from energy-efficient systems to renewable energy integration.

One significant trend is the rise of solar-powered HVAC systems. Solar energy, being abundant and renewable, presents a viable solution to power HVAC systems. Solar panels integrated with HVAC systems can harness sunlight to generate electricity, thereby reducing reliance on conventional energy sources. This not only translates to lower operational costs for businesses and homeowners but also contributes to a significant reduction in carbon footprints.

Another promising avenue is geothermal heating and cooling systems. Geothermal systems utilize the stable temperature of the earth to efficiently heat or cool indoor spaces. By tapping into this renewable energy source, HVAC systems can operate with minimal electricity consumption, thereby reducing both energy costs and environmental impact. The scalability and adaptability of geothermal systems make them a compelling option for a sustainable HVAC future.

Furthermore, energy recovery ventilation systems are gaining traction. These systems capture and reuse energy from exhaust air to precondition the incoming fresh air. By transferring heat or cold from the outgoing air to the incoming air, these systems significantly reduce the energy needed to condition the air, further amplifying energy efficiency and sustainability in HVAC operations.

Artificial Intelligence and Machine Learning

Artificial Intelligence (AI) and Machine Learning (ML) are anticipated to be at the forefront of HVAC advancements in the foreseeable future. These technologies have the potential to revolutionize the way HVAC systems are managed and operated, optimizing energy usage and enhancing user comfort.

Predictive maintenance is one of the significant applications of AI in HVAC. AI algorithms can analyze vast amounts of data collected from HVAC systems and predict when maintenance is needed. This predictive capability minimizes downtime, reduces repair costs, and ensures that the HVAC system operates at peak efficiency. Maintenance becomes proactive rather than reactive, leading to more reliable and efficient HVAC systems.

Real-time energy optimization is another area where AI and ML can make a substantial impact. These technologies can analyze data on occupancy patterns, weather forecasts, and energy prices to adjust HVAC settings in real time. For instance, when a space is unoccupied, the system can automatically adjust the temperature to reduce energy consumption. Over time, the AI learns and adapts, further optimizing energy usage and reducing operational costs.

Personalized climate control based on individual preferences is a promising application of AI in HVAC. AI algorithms can learn individual preferences for temperature, humidity, and airflow and tailor the HVAC settings accordingly. This level of customization ensures that each person's comfort is optimized, leading to a more pleasant indoor environment.

Modular and Flexible Designs

The future of HVAC design is likely to lean heavily towards modularity and flexibility. Buildings are becoming increasingly diverse, with varying architectural designs, occupancy patterns, and usage requirements. The conventional one-size-fits-all approach to HVAC system design is proving to be less effective in meeting the unique needs of different spaces.

Modular designs will be the answer to this challenge. HVAC systems will be designed in modular units that can be easily adapted and combined to suit the specific requirements of a building. This modular approach allows for a high degree of customization, ensuring that the HVAC system is efficient, cost-effective, and perfectly aligned with the building's characteristics.

Furthermore, flexible designs will enable seamless integration with other building systems. HVAC systems will be designed to work harmoniously with lighting, security, and other smart building components. This integration not only enhances overall building efficiency but also provides a cohesive and user-friendly experience for building occupants.

Electrification and Heat Pumps

As the global focus shifts towards reducing carbon emissions, electrification is set to become a dominant trend in the HVAC industry. Electrifying HVAC systems involves moving away from fossil fuel-based heating systems to electric-powered solutions. This transition is significant in mitigating the carbon footprint of buildings, which is a crucial step towards achieving sustainability goals.

Within the realm of electrification, heat pumps are poised to play a central role. Heat pumps are highly efficient HVAC systems that can provide both heating and cooling. They operate by transferring heat between the indoors and outdoors, making them energy-efficient and environmentally friendly. Heat pumps can utilize renewable energy sources, such as solar or wind energy, to power the heating and cooling processes, further aligning with sustainability objectives.

Air-source heat pumps, ground-source (geothermal) heat pumps, and water-source heat pumps are the primary types of heat pumps. Each type has its advantages and is suited for different geographic regions and building types. Air-source heat pumps, for example, are more common and efficient in moderate climates, while ground-source heat pumps excel in regions with more extreme temperature variations.

Human-Centric Approach

The future of HVAC systems will witness a paradigm shift towards a more human-centric approach. Beyond the traditional focus on temperature control and air quality, the emphasis will extend to enhancing the overall well-being and comfort of building occupants.

A key aspect of this approach is designing HVAC systems that align with human circadian rhythms. Circadian lighting, which mimics the natural light patterns throughout the day, can be integrated with HVAC systems to regulate lighting and temperature based on the time of day. This synchronization promotes a healthier sleep-wake cycle, contributing to better overall health and productivity.

Noise reduction is another essential element of a human-centric HVAC approach. HVAC systems will be designed and optimized to operate at minimal noise levels. This not only enhances comfort but is also crucial for spaces where noise can be a distraction or a hindrance to concentration, such as offices and educational institutions.

Moreover, personalized climate zones within a building will be a standard feature. Occupants will have the ability to customize the temperature and airflow in their immediate surroundings, providing a sense of control and comfort. This level of personalization caters to individual preferences and further enhances the overall indoor experience.

CHAPTER 11
Troubleshooting and Repairs

The journey of managing and maintaining various systems, machines, or devices is often met with challenges. These challenges could be in the form of unexpected malfunctions, errors, or issues that disrupt the normal functioning of the system. To effectively address these challenges, a structured approach to troubleshooting and repairs is crucial. In this chapter, we will delve into the intricacies of troubleshooting and the importance of knowing when to seek professional assistance. Additionally, we will explore real-life case studies to gain a practical understanding of the troubleshooting process.

Advanced Diagnostic Tools

Troubleshooting and repairing technical systems have become significantly efficient and accurate with the advent of advanced diagnostic tools. These tools are the cornerstone of modern maintenance and repair practices, enabling professionals to delve deeper into the intricacies of a system to identify issues and resolve them effectively. This section will explore the diverse spectrum of advanced diagnostic tools, how they function, and their critical role in the troubleshooting process.

Instrumentation for Physical Systems

In the realm of physical systems such as machinery, automobiles, or industrial equipment, advanced diagnostic tools often take the form of sophisticated instrumentation. These tools are designed to measure, analyze, and interpret physical parameters, allowing technicians and engineers to diagnose faults and malfunctions accurately.

Oscilloscopes

Oscilloscopes are fundamental instruments used in diagnosing electrical and electronic systems. They provide a graphical representation of electrical signals, enabling professionals to visualize waveforms and detect abnormalities or irregularities in the signals. This visualization is crucial for identifying issues like voltage spikes, frequency fluctuations, or distortions in the waveform that may indicate a malfunction within the system.

Multimeters

Multimeters are versatile diagnostic tools that combine several measurement functions. They can measure voltage, current, resistance, and sometimes other parameters like capacitance and frequency. Multimeters are invaluable in troubleshooting electrical circuits, checking the continuity of electrical paths, and measuring the resistance of components. For instance, they are used to diagnose issues in wiring, circuits, and electronic components.

Thermal Cameras

Thermal Cameras, also known as infrared cameras, are essential for identifying temperature variations in a system. They capture infrared radiation emitted by an object and convert it into a visible image, with different colors representing different temperatures. These cameras are extensively used in various industries to detect overheating components in machinery, electrical systems, and even buildings. An abnormal temperature reading can indicate a potential fault or inefficiency within the system.

Pressure Gauges

Pressure Gauges play a vital role in diagnosing problems related to pressure in mechanical and hydraulic systems. They measure the pressure exerted by gases or liquids and help technicians monitor and diagnose issues like leaks, clogs, or irregular pressure levels. For instance, in an automobile, a pressure gauge can reveal problems in the fuel system or the hydraulic brake system, ensuring they function optimally.

Vibration Analyzers

Vibration Analyzers are instrumental in diagnosing mechanical issues in rotating machinery. Excessive or irregular vibrations can signify misalignments, imbalances, or mechanical wear within a system. Vibration analyzers provide data on the vibration levels and frequencies, assisting in identifying and rectifying potential mechanical failures.

Software-Based Diagnostic Tools

In the digital era, software-based diagnostic tools have gained immense prominence, especially in diagnosing complex electronic and computational systems. These tools analyze software components, system logs, and various parameters related to software operations, providing valuable insights into system health and performance.

Network Analyzers

Network Analyzers are crucial in diagnosing issues within computer networks. They help analyze network traffic, identifying bottlenecks, packet loss, latency, and other network-related problems. By examining data packets and their routes, network analyzers enable IT professionals to optimize network performance and troubleshoot connectivity issues.

Diagnostic Software Suites

Diagnostic Software Suites are comprehensive tools that analyze software systems at various levels. They can identify software bugs, memory leaks, and other issues affecting the performance and stability of software applications. Developers and IT professionals use these tools to optimize code, enhance application performance, and ensure a seamless user experience.

Disk Diagnostic Software

Disk Diagnostic Software, also known as disk checking or disk utility software, is designed to detect and repair issues with hard drives and storage devices. These tools scan the disk for bad sectors, file system errors, and other disk-related problems. Disk diagnostic software can help prevent data loss and system crashes by addressing potential disk issues before they escalate.

Operating System Monitoring Tools

Operating System Monitoring Tools provide real-time insights into the performance of an operating system. They monitor various metrics such as CPU usage, memory utilization, disk activity, and network activity. Any abnormal spikes or trends in these metrics can indicate system stress or impending failures, allowing administrators to take proactive measures.

Importance of Data Analysis

One of the significant benefits of utilizing advanced diagnostic tools is the wealth of data they generate. This data, in turn, can be analyzed to gain a deeper understanding of system behavior, detect patterns, and predict potential issues.

Data analysis is crucial in identifying recurring problems. By aggregating diagnostic data over time, patterns can emerge, highlighting specific components or conditions that consistently lead to failures. Armed with this information, engineers and technicians can proactively address these recurring issues during maintenance or even redesign the system to mitigate them.

Predictive maintenance, an emerging practice, heavily relies on data analysis derived from advanced diagnostic tools. By analyzing historical data, technicians can forecast when a component is likely to fail or when maintenance is due. This approach minimizes downtime, reduces repair costs, and ensures the system's optimal performance.

When to Call a Professional

Efficient troubleshooting and repairs are vital aspects of maintaining any system, whether it's your household appliances, your car, or complex industrial machinery. While basic troubleshooting skills are essential for everyone, knowing when to call a professional can be equally crucial. Recognizing your limitations and seeking expert help at the right time not only saves time and money but can also prevent further damage or harm. In this section, we will delve deeply into various scenarios that highlight when calling a professional is the most prudent course of action.

Complex Systems and Specialized Knowledge

In today's highly advanced technological landscape, systems are becoming increasingly complex. These complexities demand a deep understanding of specialized domains. Take, for instance, the healthcare sector. Medical equipment, such as magnetic resonance imaging (MRI) machines or positron emission tomography (PET) scanners, are highly intricate and sensitive. Repairing or even diagnosing minor issues with these devices requires specialized knowledge and training that only professionals in the medical equipment field possess.

Similarly, the aviation industry presents another compelling example. Aircraft, with their intricate engineering and reliance on cutting-edge technology, require specialized aeronautical engineers and technicians to maintain and repair them. Attempting DIY repairs on an aircraft could not only be ineffective but also extremely dangerous.

Safety and Legal Concerns

Safety should always be a paramount consideration when deciding whether to call a professional. Some repairs involve handling potentially hazardous materials, high voltages, or heavy machinery. Any mistake could result in severe injury or even loss of life.

Consider electrical repairs in a residential setting. A simple task like fixing faulty wiring may seem manageable, but incorrect handling of electrical components could lead to electric shocks, fires, or even fatal accidents. Certified electricians are not only equipped with the necessary knowledge and skills but also adhere to strict safety protocols and legal requirements, ensuring the safety of individuals and properties.

In many industries, adherence to legal and safety regulations is mandatory. For instance, the construction industry follows stringent safety guidelines to prevent accidents and ensure the well-being of workers and the public. Professionals in this field are well-versed with these regulations and operate within legal boundaries to avoid potential legal repercussions.

Persistent or Recurring Issues

A recurring or persistent issue is a clear indicator that something more profound might be at play. When an issue keeps coming back despite your best efforts to fix it, it's a sign that professional expertise is required. This persistence suggests an underlying problem that requires a thorough diagnosis and solution.

For example, if your vehicle continues to have engine problems even after multiple attempts at fixing it, there might be a deeper issue that necessitates the expertise of a specialized mechanic. They can utilize advanced diagnostic tools and their experience to pinpoint the exact problem and provide a lasting solution.

Time and Resource Constraints

Time is a valuable resource, and there are instances where attempting a repair on your own could be a significant drain on it. In our fast-paced world, dedicating substantial time and effort to troubleshoot and repair a complex problem may not be feasible, especially when urgent solutions are needed.

Consider a scenario where a critical component in a manufacturing facility breaks down, halting production. Every minute of downtime incurs substantial financial losses. In such cases, calling a professional who can swiftly identify and rectify the issue is the most economical choice in the long run.

Moreover, specialized professionals often have access to resources and tools that are not readily available to the average individual. They can utilize these resources to expedite the repair process, minimizing downtime and optimizing efficiency.

Cost-Effectiveness and Long-Term Benefits

One of the misconceptions about professional assistance is the assumption that it's always more expensive than attempting a repair independently. However, this isn't always the case. While hiring a professional does involve a cost, it's often a prudent investment with long-term cost-effectiveness.

Professionals can diagnose and address issues accurately and efficiently, preventing potential damage that could escalate repair costs significantly if left unattended. Their expertise ensures a lasting solution, reducing the likelihood of recurrent problems. Ultimately, this saves you both time and money in the long haul.

Furthermore, professionals can provide valuable advice on maintenance practices that can prolong the lifespan of your equipment or system. This guidance contributes to the long-term efficiency and durability of the system, translating into cost savings over time.

Maintaining Peace of Mind

Lastly, engaging a professional for repairs can provide peace of mind. Knowing that an expert is handling the issue gives you confidence in the repair process. You can trust that the problem will be correctly identified and effectively resolved, ensuring the optimal functioning of the system.

Moreover, in situations where warranties or service agreements are in place, professional intervention maintains the validity of these agreements. Attempting DIY repairs might void warranties, causing further complications and expenses down the line.

Case Studies

In the realm of troubleshooting and repairs, case studies offer invaluable insights into the practical application of knowledge and problem-solving skills. They provide a platform to understand complex scenarios, analyze the steps taken to identify and rectify issues and derive lessons for future problem-solving endeavors. In this section, we will explore two detailed case studies that encompass different domains of troubleshooting.

Case Study 1: Network Connectivity in a Corporate Environment

Scenario

In a bustling corporate office housing hundreds of employees, a persistent issue of frequent disruptions in network connectivity began to surface. Employees across various departments reported intermittent loss of connection to the central server and a noticeable slowdown in internet speeds. The IT department initiated basic troubleshooting, but the root cause remained elusive, necessitating a deeper investigation.

Troubleshooting Approach

1. Initial Assessment:

The IT team commenced the troubleshooting process with an extensive assessment of the network infrastructure. They meticulously inspected all routers, switches, and cabling to identify any visible physical damage or loose connections that might be contributing to the connectivity issues.

2. Diagnostic Tools:

Leveraging advanced diagnostic tools, particularly network monitoring software, the IT team began monitoring and analyzing network traffic. This step was crucial in identifying congestion points, abnormal patterns, and any unusual activities that might be causing the disruptions.

3. Collaboration and Expert Consultation:

Given the complexity of the corporate network, the IT team decided to seek collaboration with network specialists. They shared the collected data and observations with these specialists, who possessed specialized knowledge and experience in dealing with intricate network setups.

4. Mitigation Strategy:

The collaborative effort resulted in the identification of specific nodes and areas within the network experiencing high traffic and congestion. Based on this analysis, a comprehensive mitigation strategy was devised. This strategy involved redistributing network traffic, upgrading certain network components, and implementing Quality of Service (QoS) settings to prioritize critical traffic.

5. Implementation and Testing:

The identified upgrades and modifications were implemented in a staged approach, ensuring minimal disruption to ongoing operations. Rigorous testing followed each stage of implementation to validate the effectiveness of the changes. Real-time monitoring during the testing phase was crucial to track improvements in network performance.

Result

The collaborative approach and specialized consultation with network experts yielded successful results. The implemented mitigation strategy led to a significant improvement in network performance. Employees reported a noticeable reduction in connectivity disruptions and improved internet speeds, subsequently enhancing their productivity. This case underscored the importance of leveraging advanced diagnostic tools and seeking expert guidance to tackle complex network issues effectively.

Case Study 2: Malfunctioning Industrial Robot in a Manufacturing Plant

Scenario

In a busy manufacturing plant focused on producing automotive components, a critical industrial robot responsible for a specific assembly line process began experiencing persistent malfunctions. The malfunctioning robot disrupted the production flow, leading to delays and impacting production targets. The maintenance team was promptly alerted to address the issue.

Troubleshooting Approach

1. Visual Inspection and Basic Diagnostics:

The maintenance team initiated the troubleshooting process with a visual inspection of the industrial robot. They checked for any visible signs of damage, loose wires, or misaligned components. Basic diagnostic tests were also conducted to determine if the issue was apparent.

2. Utilization of Advanced Diagnostic Tools:

Recognizing that the issue might be more intricate than visible signs suggested, the maintenance team employed specialized diagnostic tools. Multimeters and diagnostic software were used to analyze the electrical and mechanical systems of the robot comprehensively.

3. Consultation with Manufacturer:

Despite the initial diagnostic efforts, the exact cause of the malfunction remained elusive. Realizing the complexity of the industrial robot and the need for specialized expertise, the maintenance team sought guidance from the robot manufacturer's technical support. They shared the diagnostic data and received expert recommendations on potential causes and solutions.

4. Parts Replacement and Calibration:

Following the manufacturer's recommendations, specific components identified as potentially faulty were replaced. Additionally, the robot underwent recalibration to ensure optimal performance once the replacements were made.

5. Testing and Validation:

Rigorous testing was carried out post-repairs to validate the effectiveness of the measures taken. The robot was subjected to simulated operational conditions to ensure it met the required performance standards without any further malfunctions.

Result

The collective efforts of the maintenance team, coupled with expert guidance from the manufacturer, proved successful. The replaced components and recalibration resolved the malfunctions in the industrial robot. Production resumed seamlessly, meeting the targeted output. This case highlighted the significance of utilizing advanced diagnostic tools and seeking expert assistance, especially when dealing with complex machinery critical to production processes.

CHAPTER 12
Sustainability and HVAC

In the modern era, as we grapple with the consequences of climate change and depleting natural resources, sustainability has become a cornerstone of responsible and forward-thinking design and engineering practices. Within the realm of heating, ventilation, and air conditioning (HVAC), sustainability is not only an ethical choice but a strategic necessity. This chapter delves into the myriad aspects of sustainable HVAC practices, including environmentally friendly approaches, LEED certification, and sustainable technologies, shedding light on how these concepts intertwine to shape a greener and more sustainable future.

Environmentally Friendly Practices

As climate change and energy efficiency move front and center, the HVAC industry is embracing more sustainable design, equipment selection, and operating practices. Implementing environmentally friendly solutions significantly reduces the carbon footprint of heating, cooling, and ventilation.

At the foundation, maximizing energy efficiency curtails resource consumption while still providing adequate climate control. Utilizing high-efficiency components like variable refrigerant flow (VRF) systems, ECM motors, and boiler heat exchangers extracts more output per unit of energy input. Preventing oversizing equipment also avoids wasted capacity and needless energy usage.

Thoughtful HVAC system design further optimizes efficiency. Analyzing building characteristics like insulation levels, window efficiency, and occupancy patterns allows for right-sizing systems and components. Strategic equipment layout and duct routing minimize pressure drops and fan energy. Specifying high-performance building envelopes and passive ventilation lowers mechanical system loads.

Once installed, diligent maintenance sustains peak efficiency. Tasks like coil cleaning, filter changes, leak checks, and sensor calibration ensure optimal performance year after year. Identifying issues early prevents crises and inefficiencies. Adopting fault detection diagnostics and analytics guides proactive maintenance.

Advanced controls provide further efficiency gains. Smart thermostats with occupancy sensing reduce the heating and cooling of vacant spaces. Building automation systems (BAS) coordinate equipment sequencing, temperatures, and operating schedules holistically. Dashboards give insight into real-time performance.

Growing usage of renewable energy sources like solar, geothermal, and wind for HVAC loads decreases grid dependence. Solar panels or hot water heating supplement electricity and gas needs. Ground-source heat pumps leverage stable underground temperatures. Some systems even allow selling excess renewable energy back to utilities.

Sustainable refrigerant management is also critical. Updated regulations phase out ozone-depleting CFCs and high global warming potential HFCs in favor of options like R-32, R-1234ze, and natural refrigerants, which lessen atmospheric impacts when inevitably released. Proper refrigerant recovery during service prevents the emission of these greenhouse gases.

Adopting environmentally friendly HVAC practices does require overcoming some obstacles like higher upfront costs and design complexity. However, the long-term dividends for the climate and our communities make green HVAC an urgent investment. The industry possesses the solutions to usher in more sustainable buildings - the vision just needs purposeful implementation.

LEED Certification

Leadership in Energy and Environmental Design (LEED) is the premier program for certifying building sustainability. For HVAC systems, LEED provides a blueprint for maximizing efficiency and minimizing environmental impacts.

As major energy consumers, HVAC equipment and design play a central role in earning LEED credits and ratings. High-efficiency components that move heat and air with less electricity, gas, or water are fundamental. Load calculations that right-size systems avoid wasted capacity. Strategies like economizer cycles, geothermal loops, and passive ventilation techniques also limit HVAC resource consumption.

Indoor air quality is another critical LEED category. Advanced filtration and duct cleaning ensure healthy interior environments. Adequate fresh air ventilation dilutes pollutants generated within spaces. Monitoring and control systems maintain optimal conditions while preventing over-ventilation. The usage of low VOC-emitting construction materials further protects air purity.

Water efficiency enters systems like evaporative cooling towers, which can use substantial water if not designed properly. LEED pushes for alternative water sources, rainwater harvesting, treatment and reuse of greywater, and installation of low-flow fixtures. Even small changes accumulate across large facilities.

LEED also considers the sustainability of HVAC equipment manufacturing and material sourcing. Specifying systems utilizing low-embodied energy, recycled content, and local sourcing reduces environmental impacts before installation. Responsible equipment disposal and recycling at end-of-life further complete the sustainability lifecycle.

While energy efficiency is paramount, occupant comfort and productivity cannot be sacrificed. LEED recognizes that HVAC systems must holistically enhance building environments, not just save resources. Thermal comfort surveys and the sharing of best practices with industry groups help achieve this balance.

LEED constantly raises the bar, so innovation is encouraged. Utilizing emerging technologies like thermal storage, geothermal heat pumps, or biophilic design earns additional points. Even small creative touches to improve performance qualify.

Pursuing LEED credentials ensures that HVAC equipment and operation align with energy efficiency, conservation, sustainability, and occupant wellbeing. Environmentally conscious design, material selection, and maintenance practices translate into measurable performance gains and certifications. LEED momentum drives the HVAC industry forward.

Sustainable Technologies

Innovations in equipment and system design are key to making HVAC solutions more sustainable. Several promising technologies are gaining adoption:

Heat pumps provide highly efficient heating and cooling utilizing ambient thermal energy. Rather than burning fuels, renewable heat is harvested from outside air, ground loops, or water to warm or cool spaces. Ultra-low temperature models extend viability to cold climates. Heat pumps are a versatile, electric alternative to boilers and chillers.

Solar thermal systems harness the sun's abundant energy for water and space heating needs. Solar collectors, often mounted on roofs, preheat water for domestic use or HVAC system loops. Passive solar building designs also utilize strategic glazing to naturally heat interiors. Solar thermal reduces reliance on fossil fuel combustion.

Building integrated photovoltaics (BIPVs) seamlessly incorporate solar panels into roofs, skylights, and walls. Generating electricity on-site, they offset consumption from lighting, plug loads, and HVAC equipment. BIPVs maximize solar yield using existing building infrastructure. Combined with battery storage, surplus solar energy can power essential loads during grid outages.

Advanced control systems optimize HVAC energy use by responding to changing conditions. Smart sensors coupled with algorithms predict loads and occupancy to guide efficient operation. Dashboards provide insight into performance. Machine learning models continually improve scheduling and sequencing. Controls reduce energy waste substantially.

Thermal energy storage banks heat or cold during off-peak periods for use during costly peak demand times. Stores release cooling overnight when chilled water is most efficient to produce. Shift ice or water heating to daylight hours aligned with solar output. Storing energy when most economical and green stabilizes grid loads.

District thermal networks centrally produce steam, hot water, and chilled water at an urban scale. Distribution occurs through underground piping to connected buildings. System scale and baseload output boost efficiency through cogeneration and heat recovery. Renewables and waste heat integration further improve sustainability.

Improved ventilation components also contribute. Energy recovery ventilators (ERVs) extract heat from exhausted air, transferring it to incoming fresh air. Demand-controlled systems modulate ventilation rates based on real-time occupancy and air quality data, avoiding over-ventilation.

Material advancements are also pivotal. Low-global warming potential refrigerants cut greenhouse gas emissions from leaks and disposal. High insulation value materials with low embodied energy reduce conductive losses. Recycled and regionally sourced content decreases transportation impacts.

Adopting these technologies synergistically across new and existing buildings accelerates progress toward critical decarbonization and renewable energy goals. HVAC electrification coupled with sustainable power generation enables the heating and cooling sector to evolve beyond fossil fuel reliance.

PART V
PRACTICAL GUIDE AND RESOURCES

CHAPTER 13
Choosing an HVAC Contractor

When it comes to maintaining or upgrading your heating, ventilation, and air conditioning (HVAC) system, selecting the right contractor is paramount. Your HVAC system plays a vital role in ensuring the comfort and safety of your home, so choosing a qualified and trustworthy contractor is essential. In this chapter, we will delve deep into the process of choosing an HVAC contractor. We will explore key questions to ask potential contractors, the importance of verifying credentials, and how to obtain accurate estimates for your HVAC project.

Questions to Ask

Experience and Expertise

One of the first questions you should ask when evaluating HVAC contractors is about their experience and expertise. Understanding their background and knowledge can help you gauge their ability to handle your specific HVAC needs.

HVAC systems are complex, and a contractor with more experience is generally better equipped to handle a wide range of issues and installations. Inquire about how long the contractor has been in business and if they specialize in any particular aspect of HVAC work, such as residential, commercial, or industrial systems. A contractor who has been in the business for several years is likely to have encountered and successfully resolved a variety of HVAC challenges.

Licensing and Certification

Before you even consider hiring an HVAC contractor, it's crucial to verify that they are licensed and certified to perform HVAC work in your area. HVAC systems are intricate and potentially dangerous if not handled correctly. Hiring an unlicensed contractor can lead to costly mistakes and potential safety hazards.

Ask the contractor for their license number and check with your local licensing board or authority to ensure that it's valid and up-to-date. Additionally, inquire about any certifications or memberships in professional organizations, such as the Air Conditioning Contractors of America (ACCA) or the Heating, Refrigeration, and Air Conditioning Institute of Canada (HRAI). These affiliations often indicate a commitment to industry standards and ongoing education.

References and Reviews

To gain insights into a contractor's reputation and quality of work, request references from past clients. A reputable HVAC contractor should readily provide a list of references for you to contact. When speaking with these references, ask about their experiences, including the contractor's professionalism, timeliness, and overall satisfaction with the completed project.

Additionally, online reviews and ratings can offer valuable information about a contractor's performance. Websites like Angie's List, Yelp, and Google Reviews provide platforms for homeowners to share their experiences with local contractors. Pay attention to both positive and negative feedback, and look for patterns or recurring issues mentioned by multiple reviewers.

Insurance Coverage

Another critical question to ask potential HVAC contractors relates to their insurance coverage. HVAC work involves various risks, including accidental property damage or injuries to workers on your property. To protect yourself from potential liabilities, ensure that the contractor has liability insurance and workers' compensation coverage.

Request proof of insurance and take the time to verify its validity with the insurance provider. Don't hesitate to ask the contractor about their safety protocols and what measures they have in place to prevent accidents during the project.

Warranty and Guarantees

Understanding the warranty and guarantees offered by a contractor is essential to protect your investment. A reputable HVAC contractor should stand by their work and offer warranties on both labor and parts. Inquire about the duration of these warranties and the specific terms and conditions.

Ask for a written copy of the warranty to review carefully. Ensure it includes information about what is covered, what is not covered, and the process for making a warranty claim. A contractor who confidently offers strong warranties demonstrates their commitment to quality and customer satisfaction.

Energy Efficiency and Environmental Concerns

With increasing awareness of energy efficiency and environmental impact, it's important to discuss these topics with potential HVAC contractors. A knowledgeable contractor should be able to advise you on energy-efficient HVAC options and eco-friendly practices.

Ask about their familiarity with ENERGY STAR-rated equipment and inquire about their approach to minimizing energy consumption in your HVAC system. A contractor who values energy efficiency not only helps reduce your carbon footprint but can also save you money on utility bills in the long run.

Project Timeline and Schedule

Understanding the timeline and schedule for your HVAC project is crucial, especially if it involves significant installations or replacements. Ask the contractor for an estimated start date and projected completion date. Be sure to discuss any factors that could affect the timeline, such as weather conditions or unforeseen complications.

Additionally, inquire about the contractor's availability and how many projects they are currently managing. While a busy contractor may be in demand, you want to ensure they can dedicate adequate time and resources to your project.

Payment Terms and Pricing

Discussing payment terms and pricing is an essential step in the selection process. Ask the contractor for a detailed breakdown of the costs associated with your HVAC project. This should include the cost of labor, materials, and any additional fees or charges.

Inquire about their preferred payment methods and whether they offer financing options. Be wary of contractors who demand large upfront payments before any work begins. A reasonable payment schedule should align with project milestones, ensuring that you only pay for work as it is completed.

Communication and Accessibility

Effective communication is key to a successful HVAC project. Ask the contractor about their preferred methods of communication and how frequently they will provide updates on the project's progress. Accessibility is also crucial; you want to be able to reach the contractor or their team when you have questions or concerns.

Establish clear expectations for communication from the outset and ensure that the contractor is responsive and approachable. Good communication can help prevent misunderstandings and keep the project on track.

Maintenance and Service Agreements

Once the HVAC project is complete, your system will require ongoing maintenance and occasional repairs. Inquire about the contractor's availability for maintenance services and whether they offer service agreements.

A reputable contractor should be willing to discuss a maintenance plan tailored to your system's needs. Regular maintenance can extend the lifespan of your HVAC equipment and optimize its performance, ultimately saving you money in the long term.

Verifying Credentials

After asking the essential questions, it's time to dive into the process of verifying the credentials and qualifications of HVAC contractors. This step is crucial to ensure that you are working with a reputable and capable professional.

Licensing Verification

Verifying a contractor's license is a fundamental step in the credential verification process. Contact your local licensing board or authority to confirm the status and validity of the contractor's license. Ensure that it is current and applicable to the type of HVAC work you require.

Don't solely rely on the contractor's word or a copy of their license. Always independently verify this information to avoid potential fraud or misrepresentation.

Check for Complaints and Disciplinary Actions

In addition to verifying the license, it's wise to check for any complaints or disciplinary actions against the contractor. You can usually do this through your local consumer protection agency or the Better Business Bureau (BBB). These organizations track consumer complaints and provide ratings for businesses.

Pay attention to the nature and resolution of any complaints. While some minor complaints may not be cause for alarm, a pattern of unresolved issues or serious complaints should raise concerns about the contractor's reliability.

Insurance Validation

Confirming the contractor's insurance coverage is essential for your protection. Contact the insurance provider to verify the contractor's liability insurance and workers' compensation coverage. Ensure that the coverage amounts meet the requirements in your area.

Insurance validation is a critical step in avoiding potential legal and financial complications in case of accidents or damages during the HVAC project.

Contact References

Reach out to the references provided by the contractor. Prepare a list of questions about their experiences working with the contractor and the outcomes of their HVAC projects. Be sure to ask about the contractor's punctuality, professionalism, quality of work, and adherence to the agreed-upon terms.

References can provide valuable insights into the contractor's reliability and competence based on real-world experiences.

Research Online Reviews

In today's digital age, online reviews are a powerful tool for assessing a contractor's reputation. Explore various review platforms and read both positive and negative reviews. Look for recurring themes or consistent praise or criticism regarding the contractor's performance.

Consider the overall rating and the number of reviews to gauge the general consensus on the contractor's services.

Getting Estimates

After gathering comprehensive information about potential HVAC contractors and verifying their credentials, the next step is obtaining estimates for your project. Accurate estimates are crucial for budgeting and making informed decisions.

On-Site Evaluation

A reputable HVAC contractor will conduct an on-site evaluation of your existing HVAC system or the space where the new system will be installed. This evaluation allows the contractor to assess your specific needs accurately.

During the evaluation, the contractor should consider factors such as the size of your home or facility, insulation levels, ductwork condition, and any unique requirements. Providing a thorough evaluation ensures that the estimate is tailored to your situation.

Detailed Written Estimate

Ask potential contractors for a detailed, written estimate for the HVAC project. The estimate should itemize all the costs associated with the project, including labor, materials, permits, taxes, and any additional charges. Make sure the estimate clearly outlines what is included and what is not.

Review each aspect of the estimate carefully, and don't hesitate to ask for clarification on any items that are unclear. A transparent and detailed estimate helps you make an informed decision.

Compare Multiple Estimates

It's advisable to obtain estimates from multiple HVAC contractors to ensure you have a comprehensive understanding of the costs involved. When comparing estimates, consider not only the total cost but also the breakdown of expenses and the scope of work outlined in each estimate.

Avoid automatically choosing the lowest-priced estimate without considering the contractor's reputation, credentials, and the quality of their proposed work. Balance cost with quality to make the best choice for your HVAC project.

Ask About Potential Additional Costs

Inquire about any potential additional costs that may arise during the HVAC project. Unforeseen issues or necessary adjustments may lead to extra expenses. A reputable contractor should be transparent about potential additional costs and how they will be handled.

Understanding the possibility of additional expenses helps you budget accordingly and minimizes surprises during the project.

Clarify Payment Terms

Discuss and clarify the payment terms outlined in the estimate. Ensure you understand the payment schedule and the methods of payment accepted by the contractor. Avoid making large upfront payments before any work has begun, and be wary of contractors who demand full payment before project completion.

A fair and reasonable payment schedule should align with project milestones and provide you with peace of mind.

Review Contract Terms

Before finalizing your decision, carefully review the contract terms provided by each HVAC contractor. The contract should clearly outline the scope of work, project timeline, payment terms, warranties, and any other essential details.

Seek legal counsel if needed to ensure you fully understand the contract and its implications. Signing a well-drafted contract protects both you and the contractor by establishing clear expectations and responsibilities.

Choosing the right HVAC contractor involves thorough research, careful evaluation, and asking the right questions. From assessing their experience and credentials to obtaining accurate estimates and reviewing contract terms, each step is crucial in ensuring a successful HVAC project.

By investing time in this process and being diligent in your selection, you can find an HVAC contractor who not only meets your needs but also delivers high-quality work that enhances the comfort and efficiency of your living or working space. Your HVAC system is a significant investment, and selecting the right contractor is key to maximizing its benefits and longevity.

CHAPTER 14
Cost Considerations

In the world of construction and building management, one of the most crucial aspects to consider is cost. Every decision, from the choice of materials to the scale of the project, boils down to how much it will cost. This is particularly true when it comes to Heating, Ventilation, and Air Conditioning (HVAC) systems, which are vital for maintaining comfort, indoor air quality, and energy efficiency in any building. In this chapter, we delve deep into the various aspects of cost considerations related to HVAC systems.

Budgeting for an HVAC System

Understanding the Importance of HVAC Budgeting

Before embarking on any HVAC project, whether it's a new installation or an upgrade, it's essential to establish a well-defined budget. The budget serves as a financial roadmap, guiding the project from conception to completion. HVAC systems are not merely a one-time expense; they have ongoing operating and maintenance costs. Thus, budgeting is not just about the initial purchase but also about ensuring that the system remains efficient and effective over its lifespan.

Factors Affecting HVAC Budgeting

Several factors influence the budget required for an HVAC system. Understanding these factors is crucial for making informed decisions.

Building Size and Complexity

The size and complexity of the building play a significant role in determining the HVAC budget. Larger buildings or those with intricate layouts may require more extensive and sophisticated HVAC systems. Additionally, factors such as ceiling height, the number of rooms, and architectural features can impact the system's design and cost.

Climate and Location

The local climate and geographic location have a profound effect on HVAC requirements. In regions with extreme temperatures, HVAC systems must work harder to maintain indoor comfort. For example, heating requirements in a cold climate can significantly increase the budget. Conversely, in hot and humid areas, cooling systems may be a more substantial portion of the budget.

Energy Efficiency Goals

Energy efficiency is a critical consideration in modern HVAC design. More energy-efficient systems may have a higher upfront cost but can lead to significant long-term savings on energy bills. When budgeting, it's essential to weigh the initial investment against the potential energy savings to determine the best long-term value.

System Type and Features

HVAC systems come in various types, including split systems, rooftop units, and variable refrigerant flow (VRF) systems. Each has its own cost implications. Additionally, the features and technology incorporated into the system, such as smart controls or air purification, can impact the budget.

Regulatory Compliance

Building codes and regulations often dictate specific HVAC requirements for safety and environmental reasons. Ensuring compliance with these regulations is non-negotiable but may add to the project's budget. Moreover, some regions require environmental impact assessments or permits, which can also affect costs.

Accessibility and Installation Challenges

The ease of access to the installation site and any installation challenges can affect both the timeline and the budget. For instance, retrofitting an HVAC system into an existing building with limited access points may require additional labor and equipment, increasing costs.

Creating an HVAC Budget

Creating an HVAC budget is a meticulous process that involves several steps:

Preliminary Assessment

Begin by conducting a preliminary assessment of the project. This includes evaluating the building's size, layout, and existing HVAC infrastructure. It's also a good time to set specific project goals, such as energy efficiency targets or indoor air quality requirements.

Cost Estimation

Next, estimate the costs associated with the HVAC system itself. This includes the purchase of equipment, ductwork, and any necessary electrical work. It's crucial to obtain accurate quotes from HVAC contractors or suppliers during this phase.

Operating and Maintenance Costs

Consider the long-term costs of operating and maintaining the HVAC system. This includes energy costs, regular maintenance, and potential repairs. Energy efficiency improvements may increase the initial cost but can lead to substantial savings over time.

Contingency Planning

Budgets should always include a contingency fund to account for unexpected expenses or changes in project scope. A common rule of thumb is to allocate around 10% of the total budget to contingency.

Financing Considerations

Explore financing options, as discussed in the next section (14.2). Understanding how the project will be funded is vital for determining the budget's scope and limits.

Regular Review

Finally, the budget is not set in stone; it should be regularly reviewed and adjusted as needed. Changes in project scope, unexpected costs, or shifting priorities may necessitate budget revisions.

Cost-Saving Strategies

While it's crucial to allocate a realistic budget for an HVAC project, there are several strategies to help manage costs effectively:

Energy Efficiency Investments

Investing in energy-efficient HVAC equipment can yield long-term savings on energy bills. Although these systems may have a higher initial cost, the return on investment is often substantial.

Preventive Maintenance

Regular preventive maintenance can extend the lifespan of HVAC equipment and reduce the need for costly repairs. Including a maintenance plan in the budget can save money over the system's life.

System Zoning

Implementing zoning systems can improve energy efficiency by allowing precise control over which areas receive heating or cooling. This can be especially beneficial in larger buildings, reducing the overall HVAC load and operating costs.

Technology Integration

Incorporating smart HVAC controls and automation can optimize energy usage and reduce waste. These systems can learn user preferences and adjust settings accordingly, further enhancing efficiency.

Rebates and Tax Credits

Explore available rebates and tax credits, which we'll discuss in more detail in section 14.3. These incentives can significantly offset the initial cost of HVAC systems.

Financing Options

The Role of Financing in HVAC Projects

Financing plays a pivotal role in the execution of HVAC projects. While budgeting helps determine the total cost of the project, financing provides the means to cover these costs. There are various financing options available, each with its own advantages and considerations.

Common HVAC Financing Options

Self-Funding

Self-funding, or paying for the HVAC project with available capital, is the most straightforward financing option. It involves using existing funds or setting aside a budget for the project. While it offers control and avoids interest costs, it may strain the organization's cash flow.

HVAC Loans

HVAC loans are a common financing option offered by banks, credit unions, and online lenders. These loans are specifically tailored for HVAC projects, and their terms can vary widely. They may require a down payment and have fixed or variable interest rates.

Leasing

Leasing HVAC equipment is another popular choice. With a lease, the organization pays a monthly fee for the use of the equipment but does not own it. This option may have lower upfront costs and can be tax-deductible.

Energy-Efficiency Financing Programs

Many governments and organizations offer financing programs to encourage energy-efficient HVAC installations. These programs often come with favorable terms, reduced interest rates, or even grants to incentivize energy-saving upgrades.

Power Purchase Agreements (PPAs)

PPAs are agreements in which a third party owns, operates, and maintains the HVAC system on the property. The building owner agrees to purchase the power generated by the system at a predetermined rate. This allows the property owner to benefit from the HVAC system without the upfront costs.

Choosing the Right Financing Option

Selecting the appropriate financing option depends on various factors:

Financial Health and Creditworthiness

The organization's financial health and creditworthiness are critical considerations. Strong financials may enable self-funding, while organizations with good credit may qualify for favorable loan terms.

Project Scope and Budget

The size and scope of the HVAC project will influence the financing choice. Larger projects may necessitate financing options with more flexibility and higher borrowing limits.

Long-Term Goals

Consider the organization's long-term goals and how the HVAC investment aligns with these objectives. The financing option chosen should support the overall mission and sustainability objectives.

Cash Flow Considerations

Evaluate the impact of different financing options on cash flow. Some options, like loans, may require regular payments that need to be factored into the budget.

Tax Implications

Consult with a financial advisor to understand the tax implications of various financing options. Some options may have tax advantages that can significantly impact the overall cost of the HVAC project.

Planning for Successful HVAC Financing

Thorough Research

Before deciding on a financing option, conduct thorough research. Compare interest rates, terms, and any hidden fees. It's crucial to have a clear understanding of the financial commitment involved.

Detailed Budgeting

Align the chosen financing option with the project's budget. Ensure that the financing covers all anticipated costs, including equipment, installation, permits, and any additional expenses.

Negotiation and Communication

Don't hesitate to negotiate the terms of the financing. Work closely with lenders or lessors to secure the most favorable terms for the organization.

Compliance and Documentation

Adhere to all necessary documentation and compliance requirements associated with the chosen financing option. Proper documentation is crucial to a smooth financing process.

Rebates and Tax Credits

Rebates and tax credits provide a significant incentive for organizations to invest in energy-efficient HVAC systems. These incentives aim to promote sustainability, reduce energy consumption, and ultimately lower the carbon footprint of buildings.

Types of HVAC Rebates and Tax Credits

Federal Tax Credits

The federal government often provides tax credits for energy-efficient HVAC systems. These credits can be substantial and directly reduce the amount of federal income tax owed by the organization.

State and Local Rebates

Many states and local governments offer their own incentives, such as rebates or grants, to encourage energy-efficient HVAC installations. These incentives can vary widely based on the location and the specific program.

Utility Company Rebates

Utility companies frequently offer rebates to customers who upgrade to energy-efficient HVAC systems. These rebates help offset the initial investment and promote energy conservation.

Manufacturer Rebates

HVAC equipment manufacturers may also provide rebates to incentivize the purchase of their energy-efficient products. These rebates can further reduce the initial cost of the HVAC system.

Understanding Eligibility and Application Processes

Eligibility Criteria

Understanding the eligibility criteria for rebates and tax credits is crucial. Criteria often include the type of HVAC system, its energy efficiency ratings, and proper installation by certified professionals.

Application Procedures

Application procedures vary based on the type of incentive. Federal tax credits are typically claimed during tax filing, while state and local rebates may require a separate application process.

Incorporating Incentives into the Budget

Research and Documentation

Thoroughly research available incentives and gather all necessary documentation to ensure compliance with the requirements.

Consultation

Consult with HVAC professionals or tax advisors to understand how these incentives can be maximized and effectively integrated into the HVAC project budget.

Timing Considerations

Plan the HVAC project to align with the availability of incentives. Timing the installation to coincide with incentive programs can yield substantial savings.

Leveraging Incentives for Cost Efficiency

Cost Reduction

Incorporating rebates and tax credits can significantly reduce the upfront cost of the HVAC system, making energy-efficient options more financially viable.

Enhanced ROI

The reduced cost and increased energy efficiency resulting from incentives can lead to a faster return on investment for the HVAC project.

Environmental Benefits

By opting for energy-efficient systems through incentives, organizations contribute to a greener environment by lowering overall energy consumption and greenhouse gas emissions.

Cost considerations are paramount when planning any HVAC project. Understanding the intricacies of budgeting, financing options, and leveraging incentives like rebates and tax credits can significantly impact the project's success and long-term sustainability. It's essential for organizations to conduct thorough research, carefully assess their needs, and strategically plan to ensure that their HVAC investment aligns with their financial objectives while promoting energy efficiency and environmental responsibility.

Conclusion

As we reach the end of HVAC Bible for Beginners, it is my hope that you feel empowered by the broad yet practical knowledge gained about heating, ventilation, and air conditioning systems. While HVAC can seem intimidating at first glance, grasping the fundamental concepts opens up a world of possibilities, whether your goals are DIY repairs, boosting home efficiency, or pursuing a career in the field.

This book aims to provide a comprehensive introduction spanning HVAC fundamentals, residential applications, complex commercial systems, emerging technologies, and sustainability practices. We journeyed from the basic thermodynamic laws that govern HVAC to the intricate control systems now capable of automating comfort in large buildings. While this breadth of topics will not make you an HVAC technician overnight, it equips you with a valuable baseline to build upon through real-world experience and further specialized training.

As with any technical skill, true mastery comes from practice and repetition. Use this book as your launch pad to seek out hands-on learning opportunities. Offer to assist an HVAC technician on basic service calls to see systems up close. Use manufacturer forums to connect with other homeowners tackling DIY repairs. The practical knowledge you gain will reinforce the theories discussed here and build the confidence to take a more active role in servicing your HVAC equipment.

Looking ahead, the HVAC industry will continue innovating and pushing boundaries to meet the challenges of climate change, environmental stewardship, and energy efficiency. Much work remains in transitioning from fossil fuels to fully renewable systems and reducing HVAC's carbon footprint globally. I hope some of you feel called to be part of these efforts as designers, engineers, and innovators shaping the next generation of sustainable climate control.

Wherever your HVAC journey leads, approach it with curiosity and an eagerness to learn. Let this book be your ' Beginner's Bible' as you experience the satisfaction of optimizing the systems that keep us comfortable, productive, and safe year-round. The power to change your home lies at your fingertips - now get out there and realize the possibilities that await!

Claim Your Free Bonus

With This Qr Code

Made in the USA
Coppell, TX
08 March 2024

29901585R00057